SpringerBriefs in Environmental Science

T0213891

SpringerBriefs in Environmental Science present concise summaries of cuttingedge research and practical applications across a wide spectrum of environmental fields, with fast turnaround time to publication. Featuring compact volumes of 50 to 125 pages, the series covers a range of content from professional to academic. Monographs of new material are considered for the SpringerBriefs in Environmental Science series.

Typical topics might include: a timely report of state-of-the-art analytical techniques, a bridge between new research results, as published in journal articles and a contextual literature review, a snapshot of a hot or emerging topic, an in-depth case study or technical example, a presentation of core concepts that students must understand in order to make independent contributions, best practices or protocols to be followed, a series of short case studies/debates highlighting a specific angle.

SpringerBriefs in Environmental Science allow authors to present their ideas and readers to absorb them with minimal time investment. Both solicited and unsolicited manuscripts are considered for publication.

More information about this series at http://www.springer.com/series/8868

Frederic R. Siegel

Countering 21st Century Social-Environmental Threats to Growing Global Populations

 Springer

Frederic R. Siegel
George Washington University
Washington, DC
USA

ISSN 2191-5547   ISSN 2191-5555   (electronic)
ISBN 978-3-319-09685-8   ISBN 978-3-319-09686-5   (eBook)
DOI 10.1007/978-3-319-09686-5

Library of Congress Control Number: 2014946734

Springer Cham Heidelberg New York Dordrecht London

Printed on acid-free paper

Springer is part of Springer Science+Business Media (www.springer.com)

I dedicate this book to my grandchildren
Naomi, Coby and Noa Benveniste
and Solomon and Beatrice Gold who will
be living in this period of growing global
populations with the benefits and
disadvantages it brings. My hope is that
they and others of their generation will
contribute to the benefits side
of the worldwide social-environmental
and sustainability-development equation.
My profound thanks to the many authors
and presenters whose publications
I read and discussions I listened to that
have guided me in the writing of this book.

Frederic R. Siegel
Washington, DC
March 2014

Preface

World population will reach 7.2 billion people during 2014 with 1.3 billion in industrialized nations and 5.9 billion in the developing world, mainly in Asia and Africa. This number is estimated to grow to 8.6 billion by 2035, to 9.7 billion by 2050, and will likely stabilize at more than 10.3 billion earthlings at the end of the century. This assumes that there will not be population crashes because of lack of water and food for the growing populations in developing and less developed countries or because of killer epidemics or pandemics. It assumes as well that there will not be major wars in which weapons of mass destruction will be used and that a massive asteroid will not smash into our planet Earth. The time frame for global population growth is well within the lifetimes of grandchildren of our elder generation and of children and grandchildren of today's mid-life generation. Whether individual nations and the international community can provide the necessities of life to sustain the additional billions of souls is questionable given our existing inability to do so with our 2013 global population. About one billion persons today (~ 1 in 7) suffer chronic malnutrition from lack of enough food and good quality food with the young and elderly making up the majority of afflicted persons. At the same time, ~ 1.5 billion citizens (~ 3 in 14) lacked safe water for drinking, cooking, personal hygiene, and crop irrigation, and ~ 2 billion people were without safe sanitation. If the community of nations cannot service the needs such as those just noted, how can it hope to sustain the needs of the populations in 2035, 2050, and the more than 10 billion in 2100 if the world reaches the projected numbers.

Clearly, decisions taken to provide sustenance and a reasonable quality of life for existing and added populations will directly affect the lives of future generations. The issues of how strategists worldwide plan to deal with this population growth should be on an equal plane of concern to international planners as have been those catastrophic and persistent festering problems that captured their attention during the past decade and continue to do so into 2014 and subsequent years. Examples of these include national and global economic woes, natural or human-caused disasters, social stresses and needs, conflicts/wars/revolutions and political upheavals, pollution and diseases, and climate change, all of which impact contemporary societies and will likely affect future populations.

This book examines the complex interrelationships between the needs of a growing world population, the physical, chemical, and biological states of their environments, and evolving economic, social, and political situations. It reviews the best available methodologies that can be used now to provide partial or absolute solutions to problems that affect populations today. These and other problems will likely intensify in the future and must carry top priority status to resolve. The book also discusses research projects that are ongoing and that could help sustain people and their living environments if they prove successful. The text assesses the capabilities of global ecosystems and the wills of governmental and international agencies to apply suggested strategies to solve problems. Implementation of the best available technologies will allow them to provide the fundamentals necessary to achieve and sustain a reasonable quality of life for citizens beyond a subsistence state for what is hoped to be a stabilized world population this century.

Contents

Abbreviations

mm	Millimeter
cm	Centimeter
km	Kilometer
in	Inch
ft	Foot
kmph	Kilometers per hour
mph	Miles per hour
PSI	Pounds per square inch
^{o}C	Degrees centigrade
^{o}F	Degrees Fahrenheit

Chapter 1
Population Assessments: 2013–2050–2100: Growth, Stability, Contraction

1.1 Introduction

Global population growth continues although the annual rate of growth has decreased from a high of more than 87 million people in 1989 to 76.5 million in 2011. Statistical analyses indicate that the rate will fall to 43 million people in 2049. A dropping rate of population growth notwithstanding the world's population is projected to increase from 7.1 billion people in 2013 to more than 9.7 billion people during the mid-twenty-first century [1]. It is expected to stabilize at more than 10 billion people by the end of the century according to the United Nations Population Division, [2] the United States Bureau of the Census, [3] and other expert groups. The predicted stabilization figure represents about 40 % more global citizens than inhabited the earth at the beginning of 2013. The rate of population growth is dependent on multiple factors: (1) the number of females in or coming into child-bearing ages; (2) fertility rate or the average number of children borne by a woman in a country during her child-bearing years; (3) a worldwide increase in life expectancy; (4) government policies; and (5) religious and cultural influences. This increase in world population together with movement of large numbers of citizens from rural living to urban centers creates concerns about high population density as well as population growth [4]. The capability of the "haves" to provide for the "have-nots" is in question. This capability has not been achieved for the earth's 2013 population. It may be attained in the future with deliberative planning, adoption of technology advances, development investment, and governments that have the will to respond to fulfilling the needs of their citizens.

© The Author(s) 2015
F.R. Siegel, *Countering 21st Century Social-Environmental Threats to Growing Global Populations*, SpringerBriefs in Environmental Science, DOI 10.1007/978-3-319-09686-5_1

1.2 Reasons for the Exponential Growth in the Earth's Population

1.2.1 Human Female Population

The absolute numbers of human females in the world population coming into or in child-bearing ages (15–49 years old with 15–40 as the most productive years) increased dramatically between 1950 and 2010. Demographers calculate that the increase will continue to 2050 when their numbers are expected to decline although population expansion may reach 10.3 billion people by 2100. However, the percentage of females of child-bearing ages for the world population dropped significantly between 1950 and 2010 except for a bump for the 30–40 year age group as evident from Fig. 1.1 [1]. This decline is projected to be notably greater in 2050 with a continued drop in percentage of females of child-bearing age by 2100 is clearly evident in Fig. 1.1. The absolute number of females will continue to grow until the end of the century, but the drop in the number of children each has, mainly as a result of family planning and contraception, should result in a stabilized population. If this end number is taken as 10.3 billion people, we will learn sooner rather than later the limits of the earth's carrying capacity.

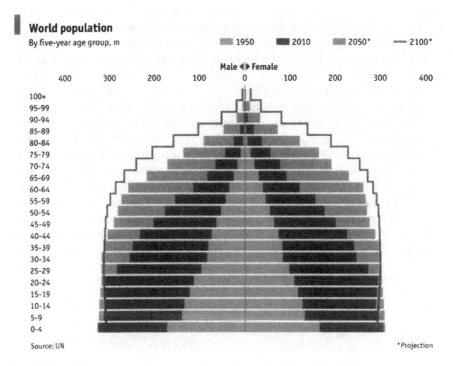

Fig. 1.1 Population by age group and sex (percentage of total population) [2]

Will the carrying capacity be able to sustain 8.6 billion souls in 2035? If so, will the sustainability extend to the 9.7 billion people in 2050? Finally, will the carrying capacity provide for a projected stabilized population in 2100? Or perhaps, we have already reached the tipping point for sustainability by not being able to provide an acceptable quality of life in 2013 for billions on our planet. This could manifest itself sometime in the next few generations by population mini-crashes in the regions least able to survive them (e.g., in sub-Saharan Africa). This need not happen if international agencies and individual countries put action plans into operation as soon as possible, not a decade or two in the future when problems may be out of hand and too difficult to cope with for most global entities. On earth, there is water to be distributed, there is unused arable land to grow food, there are natural resources to be shared, and there are social and environmental problems that can be solved. Action plans to provide for a growing world population can be implemented and ideally will bring to the fore a stable, sustainable global society. A start in activating and fully executing plans for transition as you read this would certainly be leagues ahead of awaiting population crashes.

1.2.2 Fertility

Fertility rate can be defined as the average number of children a woman has during her reproductive years (15–49, with 15–40 being the most productive). National reports to the United Nations Population Division and other organizations that monitor population data show that fertility rates in most countries have been in decline for a half century. In 1960, the global fertility rate was 4.9, and by 2011, it decreased by almost half to 2.5 but is 4.4 for the poorest countries. The 2.5 world rate in 2013 is still greater than the replacement fertility rate of ~2.06.

The decline in fertility rate is the result of multiple factors. Foremost are family planning for females and their partners and the use of contraception (mainly IUD and birth control drugs). Education and the realization that income used to provide better nutrition, health care, and schooling for fewer children is a better path than having more children who would not have these benefits. Certainly, the empowerment of women through education, earning ability, and voting and influencing government policies in many countries has been major forces in bringing down the global fertility rate.

Government policy has also worked to bring down or to stabilize fertility rates. China dictated a one child per family policy in 1979 with the result that by 2011, the Chinese fertility was 1.5, well below the replacement level. The policy applies to urban dwellers who constitute about one-third of the Chinese population but allows the two-thirds of China's population comprised of rural dwellers and ethnic groups in their autonomous regions to have more than one child per family. Birth control in China has also included abortion and forced sterilization. The Chinese population continues to increase but at a much reduced rate. This may change because of the Chinese government decision, November 2013, to allow a married couple to have two children if one of the pair was a single child, effective 2014. This affects most of the population that was born after the 1979, and one child per family policy was instituted and should result in an uptick in China's fertility rate and a rise in population.

Other examples of governments legislating to reduce their fertility rates are found in Iran and India [1]. In Iran, the law dictates that males and females must take courses on contraception before they can obtain a marriage license. This, plus socio-political, economic, and environmental concerns of families seem to work because in 2011, the Iranian fertility rate fell below replacement level to 1.9 where it remains in 2013. India has a law that allows only families with one or two children to be eligible for election to local governments where a village population is greater than 500 people. This law affects more than one billion people but is less important in bringing down the country's fertility rate than family planning, contraception, empowerment of women, and slowly breaking the tradition of having more children so as to provide security in old age. The Indian fertility rate in 2009 was 2.7 and in 2013 fell 2.4, still well above the replacement level for the world's second most populated country. By 2025, India will have the world's largest population.

Although fertility rates have fallen globally, they are still markedly high in many regions that will find it hard to impossible to sustain additional populations. For example, in 2013, sub-Saharan Africa (population of 926 million) had fertility rates of 5.2. At these rates, this African population could double in about 27 years and exceed the population of India or China. All of Africa in 2013 has 1.1 billion people, but by 2050 is projected to have 2.4 billion people [1]. However, this is unlikely as discussed in the next paragraph when the reality of problems of sustainability of additional populations and their impacts on existing populations become brutally clear. The result should be that religious and cultural barriers to reducing fertility rate would be lowered.

Whereas fertility rates greater than the replacement rate (global average 2.5 in 2013) portend badly for the earth and many national populations, the opposite situation portends badly for countries with below replacement fertility rates such as Japan (1.4 in 2013) and most nations in Europe (average of 1.6 in 2013) where populations are shrinking [1]. The continued decline in European Union countries' fertility rates is attributed by some to the 2009 economic downturn. Only France and Kosovo have rates at or above the replacement level. Shrinking populations mean that these countries, now and more so in the future, do not and will not have enough young people to fill a country's needs, whether in the service, industrial, or other sectors. Nor will there be enough young people to contribute to and sustain a county's social security net. The result is that foreign workers are recruited and accepted as a necessary fact of life (e.g., in Germany), but they are not warmly received. This is true as well for Japan with an aging population that is increasing rapidly. Cultural and religious differences, and perhaps racism, are the principal areas of contention between nationals and immigrant (foreign) workers. This can change, but it will take time and educational efforts.

1.2.3 Life Expectancy

People worldwide are living longer and today have an average life expectancy of about 70 years versus about 52 years in 1960. This is the result of several factors. Among these are improved prenatal and postnatal health care that together reduce deaths during birth, and later for babies and young children, plus more access to better staffed and stocked clinics and hospitals. When these couple with better nutrition, elimination of bad habits (smoking, drugs, excess alcohol, overeating), and taking up good habits (exercise, improved diet), all work to extend human life. Certainly, a major role is played by vaccines that basically eradicated or nearly eliminated life-threatening and crippling diseases (smallpox, polio) and reduced the incidence of others (measles, influenza). Also, the use of chemically treated nets that keep parasite-bearing mosquitos from infecting sleeping humans with malaria and dengue fever and other mosquito control programs improve health conditions and extend life expectancy in several countries. Ready communication worldwide alerts the World Health Organization (WHO) and other international, regional, and national health departments to a threat from spreading, life-threatening illnesses. Such alerts allow for the rapid transport of necessary medical personnel, medications, and vaccines to infected and at risk populations (e.g., from cholera, typhoid). This will help the sick, immunize the healthy, and in this way arrest the spread of disease, preserve life, and extend life expectancy. These factors, abetted by the facts that there are large numbers of females coming into or in child-bearing ages in developing and less developed countries, and an increasing life expectancy in these countries, explains in part why the world population will continue to increase beyond 2050. Table 1.1 gives life expectancy changes for some developed and developing countries from 1960 to 1990 to 2011 [5].

Table 1.1 Examples of life expectancy changes from 1960 to 1990 to 2011 for developed and developing countries [5]

| | 1960–2011 | | | |
	1960	1990	2011	% change
USA	70	75	79	13
Japan	68	79	83	22
Brazil	55	67	74	34
India	42	58	65	55
China	36	69	76	111
Nigeria	39	46	53	36
Russia	No data	69	69	0
South Africa	49	62	58	18*

Data for all nations at http://www.worldlifeexpectancy.com/history-of-life-expectancy
*Decline in life expectancy is the result of the HIV/AIDS epidemic

1.2.4 Obstacles to Reduction in Global Fertility Rate

Barriers to a greater global reduction in fertility rate are mainly religious and cultural (tradition). The Catholic Church has about 1.2 billion adherents or close to one-sixth of the world population. Well over half are in the Americas and Africa. The Church opposes the use of artificial contraceptive methods. This favors high fertility rates for those Catholics of the 160 million in Africa who follow the Church dogma. However, many Catholic communities in Africa may not have the wherewithal to sustain existing populations, much less added numbers, with an acceptable quality of life (e.g., access to safe water, enough food, healthcare, education, employment). Whether Church policy will change in the near future with high population growth is doubtful but possible when the reality of the earth's limited carrying capacity becomes sharply apparent as poverty, pain, and suffering increasingly afflict economically disadvantaged populations. Islam also has adherents (1.5 billion) that comprise more than one-sixth of the world's population. Islam encourages large families but does not oppose the use of modern contraceptive methods. Populous Islamic-dominated countries such as Indonesia, Malaysia, and Bangladesh have reduced their fertility rates to 2.6 or less. Conversely, the fertility rates for Somalia, Nigeria, Afghanistan, and Yemen continue high at greater than 5.5 although their rates have been falling [1].

The cultural tradition in rural areas of some countries that more children mean more hands to work the land and old age security is changing. Local educational groups and easily understood media programs work to convince parents that fewer children make education affordable and that their educated children will have opportunities to earn salaries that can better help support parents in old age.

1.3 Projections for Near-Future (One–Two Generations) Populations

If the statistical projections of population numbers for individual countries and for the world in 2013 are realized [1.5, 2.6, or 3 billion more persons in 2035 (8.6 billion), 2050 (9.7 billion), and in 2100 (10.3 billion), respectively], it is likely that these populations will exceed ecosystem earth's carrying capacity as we now assess it. In 2013, the world was not able to extend an acceptable quality of life to one billion people who suffer from chronic undernourishment (malnutrition) and to one and a half billion people who do not have access to safe water (the cause of 100s of millions of cases of diarrhea and other intestinal diseases annually). On these facts alone, we can question the global capacity to service the basic organic (biological) needs of large numbers of additional populations.

Social, economic, and political needs have to be served as well to maintain stable, sustainable societies among growing populations. Rather than assuming the worst case scenario in light of greatly expanded populations, we can seek positive solutions to obvious problems. We can strive to apply social changes, use technological advances, make economic investment, and stiffen political will in the efforts to provide an acceptable quality of life for existing populations and for future generations. Immediate global action is necessary because the future is now for today's children and those who will be their children, whether in less developed, developing, or developed nations. Delays in decision taking today will likely bring disasters in the near and far futures.

Lastly, it should be noted that statistical projections of changes in world population as each decade passes can vary dramatically as a function of data supplied by nations at the time of the analysis and the mathematical models used. Models present data as low, medium, and high growth values with most references to them using the medium figure. In an interesting exercise, projections were made by the United Nations Population Division for global population numbers until 2300 [6]. In the ensuing 2004 publication, the world population for 2050 was given as 8.9 billion people and topped out at 9.2 billion in 2075. It would then begin to decline to 8.4 billion people in 2175 and then would begin to increase to 9.0 billion by 2300. This differs considerably from the more recent figures cited earlier in this chapter that gave 9.7 billion people for the 2050 world population that was projected to top out at 10.3 billion in 2100.

1.4   Afterword

Water is the essence of life. It keeps us hydrated, it is essential to grow our food, it is necessary for personal hygiene and is basic to a sanitation system that will not cause or spread disease, and it provides a cooking medium. A human can live without food for about 40 days but without water for only about 4 days. In the following chapter, we will discuss global availability of water, how water can be managed to provide for the one and a half billion people without safe water in contemporary societies, how the water supply can be extended, and by extension, how water can be made accessible to additional billions of people that will comprise future generations.

References

1. Population Reference Bureau (2013) 2012 world population data sheet: population, health, and environment, data and estimates for the countries and regions of the world. Washington, D.C.
2. United Nations, Department of Economic and Social Affairs, Population Division (2013) World population prospects: the 2012 revision, Demographic profiles, vol II. United Nations, New York, 870 p
3. U.S. Bureau of the Census, International Data Base (2013) Current population projections: negative population growth (facts and figures). Washington, D.C.

4. United Nations, Department of Economic and Social Affairs, Population Division (2011) World urbanization prospects: the 2011 revision, New York, 50 p
5. World Life Expectancy (2012) The history of life expectancy 1960–2011. Sacramento, California. http://www.worldlifeexpectancy.com/history_of_life_expectancy. Accessed 14 Mar 2014
6. United Nations, Population Division, Department of Economic and Social Affairs (2004) World population 2300. United Nations, New York, 240 p

Chapter 2
Options to Increase Freshwater Supplies and Accessibility

2.1 Introduction

Water is the essence of life. Space programs internationally have a focus on finding evidence that at one time water in rivers/streams and oceans existed on the moon, Mars, asteroids, and other planetary orbs. The missions to these space bodies are driven by scientific curiosity, possible economic gain from processes developed in a gravity-free environment, by an evaluation of the practicability of colonization of extraterrestrial neighbors to ease the problems from overpopulation on Earth, and to assess their natural resources. The positive evidence that water has flowed on these space bodies comes from multiple observations and chemical analyses by robotic explorers. There is water on Earth to sustain a growing population if it can be accessed, if it can be kept free of pollution, and if it is allocated and managed so as to be a sustainable resource for human consumption and a necessary support for agricultural and industrial development.

2.2 The Earth's H_2O Inventory: Liquid, Solid, Gas

The growing global population will need access to water to survive and to thrive. Planners must know where the water is and develop concepts of how it can be used most effectively where it is found or where it is moved to service populations with water deficits. Finally, planners must project changes that will take place as a result of global warming/climate change in order to more efficiently use water supplies. This section will discuss how the global stocks of water are stored and the percentages of the total water on Earth each storage phase contains [1]. A later section will consider global warming/climate change as it progresses and alters the amounts of water available at Earth locations and its availability for consumption, agriculture, and industrial/manufacturing needs.

© The Author(s) 2015
F.R. Siegel, *Countering 21st Century Social-Environmental Threats to Growing Global Populations*, SpringerBriefs in Environmental Science, DOI 10.1007/978-3-319-09686-5_2

Oceans and seas contain 96.5 % of the Earth's water. It is not potable without desalinization but in a few locations seawater is piped inland and used to support an integrated project for brine shrimp aquaculture and to grow the vegetable money crop *salicornia* for European consumption [2] oceans, seas, rivers, and lakes yield fish, shellfish, kelp, and algae that are a major food source in for many global populations.

Freshwater comprises the remaining 3.5 % of water on Earth. Of this, about 1.74 % is water locked in icecaps and glaciers. Ground ice and permafrost add another 0.002 % to the inventory. Freshwater from ice is not readily available to populations other than indigenous Arctic groups that melt ice and snow as sources of the commodity. Ice is not easily transportable to where it can be harvested to produce safe water although an iceberg from Antarctica was towed by a sailing ship to Callao, Peru, at the beginning of the twentieth century to provide freshwater for a parched city. Another 0.014 % is in surface waters (freshwater lakes hold 0.007 %, rivers and streams carry 0.0002 %, and swamps have 0.0008 %). Saline lakes and inland seas contain 0.006 % of surface waters.

Subsurface waters include aquifers that contain freshwater or brackish water. Freshwater aquifers store 0.76 % of the Earth's water, 0.34 % in shallow aquifers (less than 750-m deep), and 0.42 % in deep aquifers (greater than 750-m deep). These freshwater aquifers are the important sources of water that sustain humans and the agricultural systems that provide much of our food. Soil moisture holds 0.001 % of the planet's water. Saline (brackish) water aquifers can be used for some cropping (e.g., barley) depending on water salinity and can also be feedstock for desalinization.

The atmosphere contains 0.001 % of the Earth's water. Some villages in India and elsewhere have gone from water deficient to water available for basic needs by extracting water from the atmosphere [3]. Finally, biological systems hold 0.0001 % of the Earth's water.

It is truly astounding to realize that naturally occurring freshwater that sustains most populations worldwide for potable water, for cooking, for personal hygiene, for sanitation, for irrigation and animal husbandry, and for industrial uses, represents only about 0.76 % of the Earth's total water inventory (as liquid, solid, gaseous phases). It is equally mind-boggling to learn that one and a half billion of the more than seven billion people on earth do not have access to safe water and that more than two billion people do not have water to establish sanitation systems to help them develop a healthy living condition. These are problems that do have solutions as will be discussed in a later chapter of this book.

2.3 Volumes of Water for Some Agricultural, Industrial, Commercial, and Domestic Use

Up to 70 % of the water taken from rivers and aquifers is used for agriculture to grow crops and for animal husbandry, our sources of food, whereas 22 % is used by industry and 8 % for domestic and commercial needs. The amount of water used for a given purpose varies significantly from country to country depending on the economic condition, climate, availability and efficiency of water delivery systems, and cost. For example, in the USA, 500,000 gallons of water is necessary to grow one ton of rice, whereas in Australia, the water need is 400,000 gallons, and in India and Brazil, more than one million gallons of water are necessary to grow one ton of rice. Similarly, it takes 225,000 gallons of water to grow one ton of wheat in the USA but takes 182,000 gallons in China and 625,000 gallons in Russia to do the same. The amount of water varies because each country or region has its particular seed species that is sown, climate (temperature, precipitation), hours of exposure to sunlight, topography, and soil type. Table 2.1 presents some general figures of water used in the agricultural, industrial, and domestic/commercial sectors [1, 4].

2.4 Per Capita Water Allowances in 2012 with Recalculations for 2050

There is sufficient freshwater on Earth to sustain the growing global populations. However, the distribution of Earth's water supplies is irregular geographically so that there are nations and regions with water surpluses and nations and regions with water deficits. As cited previously, in 2011, about one billion people on Earth did not have access to safe and sufficient water, and more than two billion persons did not have sufficient water to use for sanitation. Parts of Africa, the Middle East, and Asia suffer from these problems. In some African regions, for example, citizens are subsisting on far less than the 50 L/day basic water requirement (BWR) that has been set by some health organization (e.g., for drinking, cooking, personal hygiene). In developed countries, citizens use far more than the BWR for these and other purposes.

Because of population growth, the deficiencies in water supplies that existed in 2012 will intensify with time as per capita availability of water decreases, sometimes dramatically. This could lead to internal/internecine conflicts or war between nations in an effort to gain access to water especially for drinking and cropping. If there are no changes in projected future fertility rates or in renewable water supplies, the per capita water deficit condition for many countries will change from grave to critical. Table 2.2 presents examples of changes in the water availability condition that will occur in African, Middle Eastern, and Asian countries if additional water supplies are not found or accessed [3, 5–7]. Whether countries have large or small populations or have strong or weak economies, the per capita freshwater accessibility will diminish with time unless strategies are put in place to initially stabilize and subsequently increase water supplies. Some strategies are simple but would require significant investment whereas others are complex and more costly. These are discussed in the following section.

Countries with contracting populations such as Japan and many European nations with sufficient water (e.g., Germany, Italy, Russia, Ukraine) will have an

Table 2.1 Water usage by domestic, agricultural, and industrial sectors for daily living, growing comestibles, and manufacturing goods

Use/product	Water volume (in gallons)
Domestic	
Per capita home use (industrialized nations)	70
Toilet flush	1.5–3
Shower per minute	4–5
Bath	20–30
Wash dishes by hand	9–20
Dishwasher	9–12
Automatic washing machine (front/top load)	24/40
Water lawn 1 h	300
Agriculture sector	
1 ton of rice	480,000
1 ton of wheat	135,000
1 ton of corn	85,000
1 ton of alfalfa	135,000
1 ton of soybeans	160,000
1 ton of sorghum	75,000
1 ton of oats	140,000
1 ton of potatoes	60,000
1 ton of sugar beet	95,000
1 ton of sugar (from cane/beet)	28,100/33,100
1 gallon of milk	2,000
1 hen's egg	400
1 ton of beef	5,000,000 if not recycled
1 ton of cotton	2,500,000
1 barrel of beer (32 gal)	1,500
Industrial sector	
1 ton of finished steel	62,600
1 ton of synthetic rubber	110,000
1 ton of nitrate fertilizer	82,000
1 ton of paper	2,500–6,000
1 ton of fine book paper	184,000
1 ton of bricks	250–500
Refine 1 42 gallon barrel of oil	1,850
Refine 1 gallon of gasoline	63

One gallon = 3.875 L. Water use in the agricultural section is general and does not reflect the range that exists between growing areas because of climate and soil type, orientation, and other conditions [1, 8]

increasing per capita renewable freshwater supply unless a rising birth rate or climate change affects it. In theory, countries with excess water and in a favorable geographic location could export it as an economic commodity to "nearby"

Table 2.2 Growing populations will result in reduced renewable freshwater supplies per capita for most nations if population and fertility projections are met

Country	Actual renewable Water resources km³	Population Mid-2011 × 10⁶	Per capita Water 2011 m³	Population Mid-2035 × 10⁶	Per capita Water 2035 m³
Grave water stress will intensify					
Algeria	14	36	388	42	328
Burkina Faso	13	17	765	33	391
Burundi	4	10	392	20	201
Egypt	58	83	702	118	490
Israel	2	8	253	10	202
Jordan	1	7	151	9	108
Kenya	30	42	721	60	498
Libya	1	6	156	9	106
Morocco	29	32	898	39	741
Rwanda	5	11	459	20	250
Saudi Arabia	2	28	72	36	56
Tunisia	5	11	467	12	409
Yemen	4	24	168	38	104
Water stress will shift from marginal to grave					
Ethiopia	110	87	1,263	188	586
Nigeria	286	162	1,762	295	968
Somalia	14	10	1,414	17	833
Uganda	66	34	1,913	79	830
Water stress will remain marginal					
India	1,897	1,241	1,528	1,520	1,248
Iran	138	78	1,771	96	1,442
Tanzania	91	46	1,970	59	1,534
Water condition will shift from sufficient to marginal					
Afghanistan	65	32	2,006	50	1,297
Ghana	53	25	2,120	35	1,518
Water will remain sufficient					
China	2,829	1,346	2,101	1,378	2,052

This table gives examples of the degree of these reductions for some countries in Africa, the Middle East, and Asia during a generation from 2011 to 2035. Water stress is grave with <1,000 m³ per capita, marginal with >1,000–<2,000 m³ per capita, and water sufficient with >2,000 m³ per capita [3, 5–7]

locations with water deficits. Previous to the political detente, Turkey and Israel had discussions on the tanker transport of water to Israel. During December, 2013, Jordan, Israel, and the Palestinian Authority signed an agreement for the construction of a pipeline that will bring water from the Red Sea to the Dead Sea, construct a desalination plant in Aqaba, Jordan that would supply water as well to Eilat, Israel, while at the same time, Israel would release water from the Sea of Galilee (Lake Kinneret) to the West Bank citizenry.

2.5 Strategies to Improve Water Supplies

Whatever strategies are employed to improve access to renewable freshwater sources for populations inhabiting areas with water deficits will run into a barrier. This is an economics-based benefit/cost analysis of projects being evaluated: costs to build and maintain a system to provide water will exceed benefits in the short term. If such results are adjusted by factoring in social and political benefits, it will become readily apparent these will more than balance economic ones in the long term for existing and growing populations. Projects supported with this as a view toward a realistic future will likely provide the basis for important economic gains for those who supplied investment capital to bring water access strategies to fruition.

2.5.1 Import/Export Freshwater

The world has built tens of thousands of miles of pipelines to move oil and natural gas from their sources to where these fuels are used or loaded onto tankers for delivery to user nations. Investments to build the systems of pipelines were made because they transported high-priced commodities. In 2013, the price of a barrel of oil was in the US$90–100 range. The price of an equivalent amount of water in Washington, DC, with all taxes and extra charges is less than US$1 (US$0.62 a barrel equivalent). This low-value commodity, more precious to human life than oil, has not been an attractive investment for governments or organizations to build pipelines to move freshwater from where it is in excess to areas that have serious water deficiencies. This may be within a country or from one country to another. Thus, in order to supply freshwater to growing populations, a pipeline network to do so is an important strategy. Saudi Arabia has done this by laying more than 4,000 km of water distribution pipelines from Red Sea desalination plants and other freshwater sources to urban centers and industrial complexes. How easy and economically this can be done depends upon the location of a water source and a population or country that needs the water, and a government's economic capability. In the Middle East, for example, Kuwait could pay for a pipeline to bring a water supply from Iraq. If politics can be put aside, the United Arab Emirates and

Saudi Arabia have the economic capability to build pipelines to import water supplies from Iran. In Africa, the economic capability to build pipelines is limited. If international institutions would provide funds to do so, Mali could send water directly to Burkina Faso as could the Democratic Republic of Congo to Rwanda and Burundi. The cost of imported water for consumers should be gauged to local economies and earnings.

Water can also be moved via aqueducts or canals. The problem in transporting water in this way is that a significant volume of the commodity can be lost to evaporation or seepage. Nonetheless, these methods are generally used within a country. This is the case with water brought through a system of canals from the Colorado River to water-deficient Southern California, 300 mi ($\sim$480 km) away, or from Northern California to Southern California, 400 mi ($\sim$640 km) distant. Water can also be carried through tunnels excavated in mountains. This limits water loss by evaporation and/or seepage. New York City receives freshwater from a Catskill Mountains source 90 mi ($\sim$125 km) to the north via a series of tunnels. However, tunnel excavation is very costly. Thus, to maximize the volume of water received at a water poor location, transport by pipeline is preferred especially if the water is to be transported through an arid region as is the case in Saudi Arabia.

2.5.2 Find New Freshwater Sources

2.5.2.1 Aquifers

Hydrogeologists explore for confined aquifers that are not recharged from the surface and unconfined aquifers that will be recharged with freshwater as water is pumped out. Freshwater aquifers are the principal targets. However, brackish water or saline aquifers can be important in the future as stock for a desalinization project or for irrigation of crops that will grow when irrigated with brackish water (e.g., barley, asparagus). Geologists first review the rocks in an area of interest with the aim of finding sedimentary rocks such as sandstones, limestones, or conglomerates, the types that most often contain extractable water supplies. Sedimentary rocks can have porosity (voids between the grains or particles that comprise the rocks and that can store water) and permeability [interconnected voids so that water (or oil) can readily flow through the rocks]. The other two general rock types, igneous and metamorphic rocks, may have voids, fractures, and fissures that can store water and through which water can move but these are much less prevalent as aquifers. Nonetheless, they may be an important source of drinking water and must be evaluated as to their potential as aquifers. Aquifers in igneous rocks in the state of Washington in the USA provide two-thirds of the state's drinking water.

Clues in the exploration for aquifers in an area of geological interest include the presence of springs, seeps, lakes or swamps, growths of water loving trees in arid regions such as cottonwoods and willows, and topography because aquifers will be closer to the surface under valleys than beneath highlands. The geologist can also learn much from records of existing wells in a region that could help locate a new aquifer [e.g., evaluate the depth to aquifer water, the volume that can be pumped so as to maintain a steady water yield (discharge = recharge), and well records that show rocks in the strata down to the aquifer and perhaps deeper].

Once a site for exploration is selected, drilling follows. If an aquifer is found, pumpage tests are made to determine whether there is an exploitable aquifer. If so, tests continue to determine what volume of water can be withdrawn to establish an equilibrium with a recharge volume. If exploitation is indicated, chemical analyses are made on the water to assure that the quality is acceptable for drinking or other uses. An assurance of good water quality leads to exploitation in a sustainable way.

2.5.2.2 The Atmosphere

Humidity in the atmosphere can be extracted to provide additional water for drinking, cooking, and personal hygiene [8]. The process involves first moving the air through filters to remove particles, pollutants, and microorganisms. Next, the air is moved through a desiccant (silica-based gel granules) that absorbs humidity naturally. Then, water desorbs from the desiccant as steam by wind-drying, vacuum, and moderate heating. The steam condenses spontaneously at relatively low temperatures to yield clean water. Cooling speeds up the condensation process and preserves heat energy that is recycled back into the system. Minimal electrical energy is used in the extraction to condensation phases. The Indian company that manufactures water from air extraction equipment can provide additional information on the machine at *watermakerindia.com*.

The system was installed in Jalimudi village, Andhra Pradesh state in the southeastern hinterlands of India, far from pipelines, at a cost of $100,000 with a grant for electricity from the Indian government [9]. The water from air equipment provided 5 m^3 (5,000 L) of clean water daily for 600 villagers to use for drinking, cooking, and personal hygiene. The system is functional and in use in Indian factories, Rural Health Mission hospitals, dental and other offices, and homes. Larger and more costly versions of the equipment can extract up to 1,000 m^3 (1,000,000 L) of clean water daily from the atmosphere depending on the humidity. This can provide the basic water requirement of 50 L per day for 20,000 people. Clearly, the $100,000 water extraction unit could help supply water to the many small villages worldwide that do not have access to clean water or have to walk many kilometers to get water that may not be safe for human consumption. Subsidies to purchase, install, and maintain the extraction units will have to come from governments or international organizations such as the World Bank and regional development banks. In addition to India, these water extraction machines have supplied clean freshwater in the Middle East, at hospitals in Venezuela and Bolivia, where the tap water is contaminated, and to the Chinese Navy, the US Marines, and the South African Army. Other companies as well have constructed machines that extract drinking water from the atmosphere.

Another water extraction from air technique is in prototype evaluation. It operates as part of a wind turbine 34-m high that has produced 62 L an hour in a desert area near Abu Dhabi. First air is sucked into the nose of the turbine and passed through a cooling compressor behind the propellers. This extracts humidity from the air and creates moisture that condenses. Water collects and is purified as it moves through stainless steel pipes into a storage tank in the bottom of the wind turbine. The wind turbine generates the electricity that drives the water extraction operation. The system is claimed to produce 1,000 L a day. At a subsistence need of 5 L daily, the system could serve 200 people, hardly worth the projected cost of between US$660,000 and US$790,000. In addition to the cost as a problem, nature plays a role because a 15-mph wind is necessary to turn the 13-m-diameter propellers in order to generate electricity. More information can be found at www. eolewater.com/gb/our-products/range.html. The water extraction from air machine described in the previous paragraph is, at this time, the more viable method in terms of amount of water produced daily and in terms of the costs involved.

2.5.3 Recycle Wastewater into the Freshwater Inventory

2.5.3.1 Domestic Sewage

Urban populations are growing worldwide as rural dwellers move to cities for employment, education for their children, and better healthcare. In many less developed and developing countries, urban sanitation systems have not been extended to handle the wastes from this added population and the wastes are disposed of in open canals that discharge into rivers or oceans or onto land. This presents a health hazard that must be dealt with. Urban centers that already suffer from water deficits have an added burden on water supplies from an increased population. Both problems can be solved with investment in pipe-based systems that collect the wastes and carry them to a sewage treatment plant. Here, the solids are separated from the liquids. The liquids are treated to remove pathogens, put through a chemical treatment that purifies the waters, and are then ready for distribution through another system of pipes to users either at a series of fountain discharges or at dwellings.

The costs of building a collection–treatment–distribution facility and maintaining it are high but so are the sociopolitical and economic benefits that keep a population healthy without contracting illnesses bought on by poor sanitation. This allows citizens to work at their jobs without losing productivity because of absentee sick days. Although the cost of building and maintaining such a facility is high so are the short- and long-term benefits. As with other investments that increase water supplies, costs for the construction of such facilities must be borne or shared by taxpayers, governments, and by international organizations such as the World Bank, the African Investment Bank, the Inter-American Development Bank, and

US Agency For International Development. Non-Governmental Organizations that have strong financial resources share in sanitation/water availability projects.

2.5.3.2 Wastewater from Commercial Ventures

Domestic and commercial water use combined comprises 8 % of consumable water now used on Earth. The commercial wastewaters originate from many sources such as hotels and resorts, restaurants, office buildings, business parks, large condominiums, and cooperative complexes. Together with these sources, hospitals, schools (K-12) and colleges/universities, government and military facilities, retail sales stores in an out of malls, laundries, and car washes comprise a true a multitude of sources with a myriad of uses for water.

Water to service these needs is drawn from aquifers, from surface waters that likely have gone through a treatment facility before being distributed, or is delivered by water suppliers. Much of the water is used to flush toilets and other sanitation needs as keeping commercial sites clean (e.g., for healthcare), or as in food preparation. However, depending on the climate (e.g., in a warm region), large volumes of water can be used for air-conditioning and in (esthetic) lawn maintenance. For many commercial operations that exist in developed countries, in urban areas of developing countries, and in some population centers in less developed countries, the wastewater flows into a collection facility where it is subjected to treatment before being recycled or released into surface waters. Problems arise when collection and treatment facilities do not exist so that contaminated wastewater is released into open canals that run through populated areas, thereby creating human health hazards along their flow paths into streams, rivers, or oceans.

2.5.3.3 Industrial Wastewater

Of the 22 % of the Earth's freshwater inventory used by industry, about 13 % is used in power generation and about 9 % for all other industrial ventures. About 59 % of the 22 % freshwater supply is used in high-income countries whereas only about 8 % of this supply is used in low-income countries although this latter figure is increasing markedly. Most industries supply their own water, mainly from surface sources with the rest from groundwater. Others purchase it from city water supplies (wells, collection–treatment plants). Lack of a reliable supply of useable water constrains industrial development. The demand for more water by industry is rising because of a growing demand for consumer products in developing and less developed nations and because of the increased number of factories in these countries. This latter increase is driven by two forces. One is internal investment and development as a national plan. The second is investment by industries from developed nations moving factories to low-income countries as the result of one or a combination of factors. The most important factor is the availability of reliable water and electricity supplies. Without this, there would be constraints on industrial development. Others include government incentives (e.g., tax relief), an educated and trained labor pool, lower labor costs and less union influence, and less stringent environmental restrictions for the treatment and discharge of wastewater. The relocation of industrial ventures to low-income countries is often close to urban centers. If wastewater from factories in or close to these centers is not collected, treated, and recycled, but is simply discharged into the immediate or nearby environments, it will likely harm people and damage ecosystems.

Major industrial uses of water are for the production of food, chemicals, paper and paper products, primary metals, and gasoline, and oils. The water is used for product fabrication and may be incorporated into the products. It is used for processing, washing, diluting, and cooling during industrial operations, and for transport of products. Industrial facilities with a small number of employees or large ones with 100s or 1,000s of employees use water for drinking and for washing and flushing. Industrial contamination of water compared with pollutants from other sources (e.g., domestic, commercial, agricultural) is generally more toxic (e.g., with organic and inorganic chemicals including heavy metals, toxic sludge and solvents), more concentrated, harder to treat to remove the toxins, and longer lasting when insufficiently treated wastewater invades an environment.

The amount of industrial wastewater that is collected, treated, and recycled is calculated to be 30 % of the total [10]. The other 70 %, estimated at 300–500 million tons of untreated industrial waste, are released onto land where they may pollute soils and/or infiltrate through soil and rock to pollute aquifers, or flow into streams, rivers, and oceans where they may damage ecosystems that otherwise support life and are productive. Collection and treatment of industrial wastewater to a quality that can be recycled into a source industry process will help sustain a water supply by decreasing water withdrawal needs. Japanese industries recycle 90 % of their wastewater and find that higher productivity and increased profits are associated with greater reuse of treated wastewater. The industrial discharge of treated wastewater is essential to keep ecosystems healthy and productive for humans and other living things.

The quality of water needed for different industrial processes varies so that wastewater treatment is tailored to an industry's requirements. For example, pharmaceutical and high-tech (e.g., computer components manufacturing) industries require very high-quality water as does food processing. Industries that are able to use a lesser quality water have lesser treatment expenses to recycle and reclaim their wastewaters unless they intend to release it into the environment. It is a long-term financial investment to build the infrastructure to manage the collection and treatment of industrial wastewaters and the redistribution of clean freshwater. However, as already noted, the long-term benefits will equal or likely exceed the cost with improved profits for investors. Thus, each industry requires its own specific treatment to generate the quality of water it needs to function or to discharge clean freshwater offsite. Technology advances in treatment protocols will reduce costs and bring down barriers to the use of collection, treatment, and recycling of wastewater to yield toxin-free water or allow its discharge into ecosystems.

There is no question that industry can use water in a sustainable manner and maintain a high level of productivity. Industrial development can take place without harming the environment and diminishing productivity. The keys to achieve this result are the management of water supplies with the goal of decreasing the volume of water withdrawal and by supporting research that generates innovations in wastewater treatment to quality levels necessary to an industry or for discharge into the environment. The implementation of an installation and use of these innovations as they are tested and proven reliable will yield cleaner processes, better products, and lead to a greater degree of sustainability. This management of water supplies and the drive to a greater degree of recycling is basic to the ability to adapt to times of water scarcity that are projected to be more pronounced in many areas (e.g., Africa, Australia, South America) as global warming causes changes in water supplies.

The principal competition for existing freshwater supplies is from the agricultural sector, and this will have to increase in the future for two main reasons. The first is population growth and the need to grow more crops to feed more people (1.4 billion more by 2035 and an additional more than a billion people by 2050). The second is that more grain and fodder will be needed to feed livestock because there are more people with disposable income in developing countries (e.g., China) and to some degree less developed countries that are altering their diets to include more protein (e.g., meats) similar to populations in developed countries. Much of this dietary change is the result of more populous middle classes that travelled abroad and experienced diverse foods they enjoyed but with which they had not been familiar. The change is also stimulated by media exposure especially via television and Internet sites that have a constant flow of food-related propaganda. The problems of reclaiming and recycling of wastewater from agricultural production are discussed in the following section.

2.5.3.4 Agricultural Wastewater

As noted at the beginning of this section, 70 % of freshwater resources are used in the agricultural sector, mainly for irrigation where rain fed agriculture is not reliable for cropping because of a variable climate. Irrigated fields are most important because the 20 % of the world's agricultural land now farmed using irrigation produces 40 % of the global crop output. Large amounts of water are used as well in livestock production for drinking, cooling, and cleaning. The volume of water used in livestock farming is increasing greatly as the demand for livestock products (especially beef) rises with an improving economic status of a growing middle class of citizens in developing and less developed nations. Questions we deal with here are where in the agricultural sector, is it possible to reclaim wastewater and how much of this water can be economically reclaimed and reused? If not reclaimed, what then? We are also concerned with minimizing drainage of wastewater that flows overland and contains or incorporates nutrient (e.g., potassium K, nitrogen N, phosphorus P) or toxins (e.g., arsenic As, pathogens). These can infiltrate aquifers or discharge into fresh, brackish, and saline surface waters, disrupting the health condition and productivity of ecosystems they feed. Finally, there is the problem of treating water that contains wastes generated by aquaculture and recirculating it into manufactured fish farm aquaria.

Waters used in irrigation reenter the hydrological cycle and are naturally "scrubbed" as they recycle into aquifers and surface waters for reuse. As populations continue to grow, the mass of food that will have to be produced will rise dramatically. Estimates are that by 2050, food production will have to increase by 70 % (or more, author's determination) to feed the estimated additional 2.5 billion or more people added to the 2013 global population that today has one in seven people on Earth suffering from malnutrition because of the lack of access to enough and good quality food [11]. For this reason, more water will be needed by the agricultural sector even as hybridization and/or genetic engineering develop seeds for crops that require less water to grow or that improve their drought resistance. For rain fed cropping, this means developing the capability to store captured rainfall in reservoirs (that can lose a great volume of water to evaporation) or in underground caverns as backups to irrigate rainwater-starved crops.

Human-directed recycling in the agricultural sector has relatively few and somewhat localized or regionally important targets unlike targets for the recycling of industrial waters. The agricultural targets are associated with livestock production, slaughter, and preparation of products. This could be considered agri-industrial that perhaps could just as well have been discussed previously in the industrial section.

The larger targets for reclamation and recycling of wastewaters include cattle feedlots containing thousands of cattle. The Simplot feedlot outside of Grandview, Idaho, USA, has a capacity for 150,000 cattle. The USA had 10.5 million head of cattle cycling through feedlots in 2010. Commercial chicken farms for meat or eggs (some with tens of thousands and up to a million chickens) present sites for recycling of wastewater. Smaller but no less important areas for collecting wastewater, treating them, and reusing the clean water include slaughter houses, meat-packing plants, especially those that have to process 500,000 head of cattle annually to be profitable, and chicken-processing plants. Intensive farming of pigs, ducks, turkey, and geese add to the draw on freshwater. These animal husbandry sites are focused for reclamation of wastewater. Aquaculture farms are also locations where water can be cleaned and recycled perhaps following fish tank water reclamation methodology. Other candidates for water collection, treatment, and reuse are farms with dairy cattle. Factories that process fish from marine and saline or freshwater catches are also sites where wastewater can be captured, treated, and reclaimed. Recycling in all these cases will reduce withdrawal of freshwater from aquifers and surface waters, thereby improving water supplies. Treating wastewater will help secure the cleanliness of surface waters and aquifers.

Cattle feedlots in the USA are categorized as small when they contain less than 1,000 head of cattle, as medium-capacity lots when the number of cattle is more than 1,000 but less than 31,999, and of large capacity when they feed more than 32,000 head of cattle. A feedlot in Broken Bow, Nebraska, contains 85,000–90,000 cattle. Smaller feedlots comprise 95 % of those in the USA but 80–90 % of the fed cattle are in the larger feedlots. Cattle are generally brought from grazing land to feedlots where they are confined for 4–5 months until their weights increase from 600–800 lbs (273–364 kg) to 1,000–1,250 lbs (455–545 kg), and they are ready for the slaughterhouse. The amounts of waste that is produced are great. Each head of cattle in feedlots creates more than 15 times the waste daily that a human produces. Thus, a lot with 10,000 head of cattle originates the same amount of waste as a city with 150,000 inhabitants, and the Nebraska feedlot with 85,000 cattle produces wastes equal to that produced by a metropolis of about 1–1/4 million inhabitants. Feedlot wastes have been found in surface waters in watersheds far from a point source and in aquifers.

In this section, we are dealing with water that cattle drink, water used to cool and clean feedlot stock, and urine, and whether these fluids can be recycled as water cattle can safely drink. The discussion on cattle that follows applies as well to commercial farming of pigs. Cattle are penned in at feedlots that are constructed so that urine and other wastewater will runoff into collection/catch drains through a system that allows sediments to deposit. It then flows into retention ponds for treatment. This prevents pollution of streams, rivers, and lakes by nutrient-rich (potassium, phosphorus, nitrogen) wastewater. The holding ponds should be located away from recharge areas for unconfined aquifers. To prevent pollution of aquifers, the holding ponds should be constructed with impermeable liners such as clay or bioresistant plastic. In addition, the lined ponds must have secondary low areas, also lined, to catch and retain any overflow from the ponds during an extreme weather event without endangering adjacent land surfaces, aquifers, streams, rivers, or lakes. If a management system is in place to move feedlot wastewater from the collection sites (the holding ponds) to treatment plants, the cleansed water can be recycled for feedlot reuse and reduce the need to withdraw much additional groundwater or surface water. Water management systems have had some success but in the USA, for example, drainage from all livestock into freshwater resources still accounts for 33 % of phosphorus and 32 % of nitrogen loading, 37 % of pesticide loading, and for 50 % of the influx of antibiotics into catch basin waters [12].

Meat-packing plants are sources of wastewater and cattle body fluids that can contaminate aquifers and surface waters if they are not collected and treated to remove contaminants before reusing the water and fluids or discharging them into the environment. This problem is serious worldwide. For example, estimates for 2013 have the US plants producing 11 million tons of beef annually with Brazil, the European Union, and China, having annual productions of 9.9, 7.8, and 5.8 million metric tons of beef, respectively [13]. These figures represents about a 60 % increase during the first decade of this century. Cleanliness at these cattle processing facilities and the treatment of the waters or fluids they discharge to remove inorganic and organic pollutants is essential lest there be public health problems that affect societies and environments.

In 2009, there were an estimated 50-billion chickens (plus ducks, turkeys, and geese) in commercial factory farms worldwide, 9 billion in the USA, 7+ billion in China, and 6 billion in the European Union. [14]. Brazil and Indonesia also had high poultry populations. The emission of urine and its capture and treatment for recycling clean water is not a problem for commercial poultry factory farms because birds (poultry) do not have urinary bladders and hence do not issue urine. The urine from the poultry kidneys is continually added to digested feed that results in the dehydration and precipitation of uric acid precipitates. This is the white matter that comprises poultry droppings. The water that can be reclaimed and recycled for chicken raising is the water used to clean chicken pens after each 5–6-week growth period when the chickens reach the target 3–3½ lb weight (~ 1.5 kg) before they are slaughtered and dressed for sale. This water contains nutrients, antibiotics, pesticides (some with arsenic), vaccines, and other chemicals used to keep the poultry healthy and growing, in addition to cleaning compounds that are used. These waters must be collected and treated to remove contaminants before they recycle into a poultry farm or are discharged onto terrain or into surface water bodies. Failure to do so will pollute soils, aquifers, rivers, streams, and lakes. Similarly, water used to wash and clean chickens during the dressing stage and water used to clean the dressing areas carry these same contaminants and has to be captured and treated in the same way before recycling or discharge. What applies to commercial chicken farming applies as well to commercial duck, turkey, and geese farming.

Chickens in commercial egg-producing farms present the same water contamination problems and solutions as those used at chicken farms to produce meat. One farm in California, USA, has more than 800,000 laying hens. These require about 320,000 L ($\sim 80,000$ gallons) of water daily for the hens. Reclamation of the water used in cleaning the hen houses can be recycled into the farm water supply. Different phases of chicken farming (in the tropics) whether for meat or eggs, and of the farm as a commercial enterprise have been published by Agromisa, a Netherlands firm [15, 16].

2.5.4 Create/Extend Freshwater

2.5.4.1 Desalination

Nations with marine coastal zones and a deficit in freshwater supplies can generate freshwater by desalination of seawater. However, desalination is an energy consuming, hence an expensive process, but one that can provide a reliable supply of freshwater. Saudi Arabia is an oil-rich state in a water poor region. The country generates about 50 % of its municipal freshwater needs and 70 % of its drinking water requirements by desalination of Red Seawater using its oil to generate the energy needed by its desalination plants.

There are about 15,000 desalination plants worldwide that provide freshwater to areas that have natural water deficits. Another 120 are in a planning or construction phase. Most draw their feed water from oceans/seas. The biggest desalination facilities are in the Middle East in Saudi Arabia, the United Arab Emirates, and Israel. The Raz Azzour and the Shoaiba 3 plants in Saudi Arabia produce over one million m^3 daily while the Ashkelon plant in Israel produces close to a half a million m^3 daily. The largest inland facility in El Paso, Texas, USA, uses brackish groundwater as the feedstock for desalination and produces 105,000 m^3 daily. Most desalination plants produce less than 5,000 m^3 daily. The 2010 production of freshwater from desalination was more than 68 million m^3 and is projected to grow to 120 million m^3 by about 2020 although this figure may be too high as funding slowed because of the global recession in 2008.

More than 85 % of desalination plants use either multistage flash distillation or reverse osmosis. Three quarters of these are in the Middle East. The Raz Azzour and the Shoaiba 3 plants cited above use the multistage flash process while the Ashkelon plant uses reverse osmosis. Reverse osmosis uses less energy than thermal distillation and is less expensive. It is the fastest growing distillation technique. About half of the existing desalination plants use reverse osmosis. Desalination plants using either method bring the 3.5 % salt content of seawater (higher for the Red Sea or Mediterranean Sea) down to 0.05 %. The output volume of freshwater is about 60 % of the seawater input. This means that the brine issuing forth from desalination processes has a high salt content and poses a waste disposal problem that will be discussed at the end of this section.

The multiflash stage distillation technique uses seawater as the feedstock at pressures less than atmospheric pressure so that the seawater boils at less than 160 °F. At this temperature, scaling (buildup of foulant salts) in the equipment is greatly reduced. The system is designed so that the seawater passes through progressively lower pressure conditions and steam flashes off at steadily decreasing boiling temperatures and is captured. This technique is energy intensive. It is used in most plants in oil-producing countries in the Middle East (e.g., Raz Azzour and Shoaiba 3 cited above) because the oil is available at a much lower cost than would be the case in non-oil-producing areas.

For the reverse osmosis method, seawater (or wastewater from domestic, commercial, industrial/manufacturing sources) is pressure driven through semi-permeable membranes that are manufactured from polymers of polyamide plastics. These separate salts from water. The pressure ranges from 800 to 1,200 PSI (pounds per square inch) for seawater and from 250 to 400 PSI for brackish water depending on the specification of the membranes used. The membranes are tailored specific to the characteristics of the seawater or wastewater feed such as salinity, contained particles, chemicals, and organic contents. In some plants, dissolved matter is separated from the water by sequential passes through membranes with decreasing size characteristics. This and a final purification pass through nanofilters that can remove bacteria, viruses, pesticides, and herbicides and give the desired freshwater product.

An innovative large-scale desalination facility that is projected to provide up to 30,000 m^3 daily is being built in Al-Khafji, Saudi Arabia, a city of 100,000 inhabitants. The plant will use renewable solar energy as its power source [17]. Solar power will become increasingly important in Saudi Arabia where there is an average of 7–11 h of sunshine daily, an average temperature of 25.3 °C, a monthly range of 19 °C, and a high of 45 °C. The solar energy electricity will be generated by IBM-developed ultra-high concentrator photovoltaic cells (a lens focuses the sun's rays on the cells) to drive its seawater desalination operation and a new water filtration technology (specially designed membranes) in its reverse osmosis process. In a second phase, the plans are to ramp up the freshwater production to 100,000 m^3 per day. Electricity storage capability and/or backup electricity generating systems, likely fueled by natural gas, have to be in place to keep the plant functioning during days when there is no sun to power the photovoltaic cells. Saudi Arabia is using about 1.5 million barrels of oil a day to power the government run 30 desalination plants. Though the real cost of supplying oil to the desalination plants from the Saudi fields is very much less than the world price per barrel, the amount still represents an economic burden. By taking advantage of solar power and advanced filtration techniques in new installations, and by retrofitting existing plants to function as hybrids, desalination costs to provide more than 50 % of the country's municipal water supply and 70 % of its drinking water requirements can be reduced significantly [18].

Electrodialysis is one of the earlier methods to produce freshwater from saline waters. It is a membrane technique in which an electrical voltage is used to drive dissolved salts through a series of alternating charge-selective membranes that allow either positive or negative charged ions to pass through leaving a less saline water as a product. An anion-selective and a cation-selective membrane are coupled to make a cell. The cells are grouped in threes with the two outer cells passing the brine or seawater solution with the central cell carrying the dilute solution or freshwater. Several hundreds of cells may be stacked together to form an electrodialysis desalination system. The equipment had a problem with scaling. This was overcome in a reengineered electrodialysis reversal system that flushes scales and other contaminants off the membranes by adding a self-cleaning phase during which the polarity of the voltage is reversed several times an hour. The electrodialysis reversal unit results in higher recovery and longer membrane life.

Freshwater is also being generated in relatively small volumes to serve personnel on nuclear-powered naval vessels and at nuclear power facilities by using excess heat to drive the thermal distillation process (e.g., in Russia, in South Africa, and in Japan until the 2011, tsunami caused a nuclear plant accident and the subsequent shut down of all Japanese nuclear power facilities).

The brine wastewater discharge from desalination plants presents a special challenge because of the concentrated salts content (~ 8.6 %) that represents 86 kg (189 lbs) per metric ton of wastewater. Added to this is the problem of the chemicals used in descaling matter deposited during desalination, and the temperature of the wastewater. Can it be used? Can it be released directly into the environment from which it was taken? If so, why? If not, why not? Can the brine itself be recycled for use in another sector? One possibility depends on the industrial infrastructure at sites not too distant from a desalination plant. The brine outflow can be directed to drying ponds and the dried salts can be sold to a chemical industry that can process them to extract chemical elements and compounds that can be sold to for industrial/manufacturing use. These could include purified salt (NaCl) as condiment, salts to melts snow/ice, potassium (K) for use in fertilizer, or bromine (Br) for pharmaceuticals. Disposal of the chemical residue has to be controlled and enforced by environmental regulations.

Impacts of desalination plant discharges on the marine environment have been reviewed critically [19]. Researchers emphasize that a desalination plant brine outflow, warm and with a salinity at ~8.6 % should best be cooled, diluted (e.g., with seawater), and only then released into the ocean. It can also be discharged, cooled, but without dilution directly into the marine environment with outfalls a good distance away from unique, biologically diverse coastal ecosystems. Closer to such environments, the discharge can affect life forms in the intruded ecosystems to a greater or lesser degree depending on the wave, tidal, and current activity at the discharge zone. High-energy turbulent coasts with continual flushing will suffer less environmental impacts, and these may be within tens of meters of the discharge outfall. In older multiflash stage plants, some outfalls were in quiet coastal environments. These underwent widespread alterations to community structure in sea grass, coral reef, and soft sediment ecosystems. Discharge that has minimum effects on marine ecosystems would be in areas 100 s of meters offshore. In all cases, before–after control monitoring of the discharge outfall zones ecosystems is an absolute necessity.

An excellent review of the technology, political, economic, and social factors involved with desalination was published by the US National Research Council [20].

2.5.4.2 Nanofiltration of Tainted River Water

The French water authority designated Vivendi/Generale des Eaux to work with The Dow Chemical Co. to develop a membrane tailored to remove pesticides, herbicides, bacteria, and viruses from the Val d'Oise River water, yet leave in place dissolved minerals important to human nutrition and thus convert the tainted water to potable water [21]. Traditional purification methods did not work on the Oise River water. The area needed a new source of potable water to reduce the withdrawal of groundwater that was lowering the aquifer water table and resulting in land subsidence. The NF200 membrane was developed to remove molecules in the 0.001-μm-size range (1/10,000th of the thickness of human hair) and with a molecular weight of 200. This nanofilter does not remove heavy metals [22]. Two problems engineers had to solve was fouling by biological matter of microbial origin at the nanofilter surfaces and scaling by inorganic foulants. They did this by incorporating an automated system into the process that used anti-scalants and other cleaning agents.

A pilot study to provide clean, safe water to 5,000 people living near the river was carried out and was successful [23]. The nanofiltration system was ramped up and by 2011, 9120 of the specially designed and built membranes were installed at the Mery-sur-Oise water treatment plant. Here, the polluted water is decanted and chemically treated for a few days to allow sedimentation of solids and precipitates. The water is then subjected to ozonization followed by filtration through sand and lastly through charcoal to further cleanse it. In a final stage, the water is (low) pressure driven at 8–15 bars through spiraling tubes of the NF200 nanofilters that present more than three million ft^2 (278,709 m^2) of filter surface. This system is supplying about 140,000 m^3 (37 million gallons per day) of potable water to about 500,000 people ($\sim$ 300,000 households) located just north of Paris. Capital investment for this plant was less than 200,000 euros. The cost of production is about 0.10 € (US$0.13) more than that of a conventional treatment plant. Nano-filtration to cleanse contaminated river water where traditional water treatment is not effective requires that the nanofilters be designed and manufactured according to the physical, chemical, and biological characteristics of the river (feed) water.

2.5.4.3 Reduction of Water Use for Specific Operations

In addition to recycling of reclaimed freshwater from the various sources discussed in this section, a reduction in the use of freshwater can improve the per capita allowance of freshwater for stressed populations as their numbers grow. Drip irrigation targeted to individual plants in a field with the amount of water necessary for optimum growth is one method in use by many agriculturalists in water-stressed regions. This reduces the volume of irrigated water delivered by other methods. By using less water to produce crops that have excellent yields and nutritional value, agriculturalists benefit economically by lowering water supply costs. The use of (DNA) marker-assisted selection (MAS) to produce seeds, following natural hybridization methods that can grow basic food crops with less irrigation water input has also reduced water needs. For example, in the 1970s, one million gallons of water were used to grow one ton of rice. By the end of the century, this had been reduced to less than 500,000 gallons. As noted previously, Australian rice cropping is most efficient and uses less than 300,000 gallons of water to grow one ton of rice. Clearly, there is a range of water use for the same crop yields that exists from growing area to growing area because of climate and soil type conditions. In addition to traditional hybridization, marker-assisted or not, genetic engineering or manipulation of traits from one species to a different species aims to reduce water needs while maintaining or improving crop yields and nutritional values. However, genetically modified organisms are not accepted by countries in the European Union and others outside the Union (mainly in Africa) so that water use to grow a crop may not be reduced there. As noted earlier in this chapter, the agricultural sector uses 70 % of the Earth's freshwater. It is the agricultural sector that has the best potential to reduce the volume of freshwater it uses, thus making more available to expanding populations.

2.6 Benefits/Costs of Improved Water Supply

The costs of improving a freshwater supply vary greatly. For example, generating freshwater from seawater, brackish groundwater, and wastewater depends on the volume of water to be processed (or how much output is planned for), location with respect to the feed water (e.g., depth to aquifer for brackish water), the quality of the water to be delivered, and where clean water is to be sent. Expenditures include the capital cost of the construction of the processing installation and infrastructure, and the operating and maintenance costs. These vary greatly for whichever system is used to produce clean water. For example, the capital cost for a desalination plant and infrastructure can be as little as US$115 million for a small installation or as much as US$1.6 billion (the cost of the first-phase Shoaiba plant and subsequent upgrade in Saudi Arabia). Operating and maintenance costs range from US$2–$11 per 3.75 m^3 of water. The Shoaiba plant produces over 473,000 m^3 daily. The Kwinawa plant in Perth, Australia, had a capital cost of A$387 million and yearly operating and maintenance costs of A$24 million to produce more than 120,000 m^3 per day. Benefits of more water that is accessible to more people are great if it keeps them alive and healthy, able to work and contribute to the well being of a family and a nation in a socially stable society with human, civil, property, religious, and political rights. In the long term, these benefits on which we may not be able to put a monetary value will exceed the costs of making freshwater available for people, agricultural endeavors, and industrial and manufacturing projects. Without the benefits of systems that provide safe water to growing populations, socioeconomic expenses will be enormous because of the economic and political problems that will develop, diseases that will punish societies, and the conflicts and wars that will surely ensue to control water rights.

2.6.1 Support Research

Continued research on how to conserve water, more efficiently recycles water in industrial, manufacturing, and farming operations, collect and treat contaminated water, distribute clean water, and create freshwater that has to be supported by international organizations and governments. Researchers have to evaluate the influence of global warming/climate change on the hydrological cycle and changes in water supplies between regions. This will provide governments the basis on which to plan adaptation strategies that will temper the evolving effects of shortfalls in water supplies for populations and the ecosystems that sustain them through all sectors that support societies and their quality of life.

2.7 Afterword

Water is a principle factor in sustaining a reliable food supply for global citizenry. With this factor positive, we can examine what can be done to increase the world's food stocks to be able to feed a growing global population that is projected to put 2.6 billion more people on Earth in less than two generations by the year 2050. This is largely a subject that agronomists and plant scientists have to solve not only for today but for a future that will bring about more changes in climate than we observe and measure at the beginning of the second decade of the twenty-first century. National and international social, political, and economic decisions must be decided with this in mind. The next chapter will discuss how today's agriculturists can increase global food supplies and how governments must agree to free movement of food from where it is grown or stored to where it is needed to provide for people in regions with food deficits.

References

1. Siegel FR (2008) Demands of expanding populations and development planning—clean air, safe water, fertile soils. Springer, Berlin, Heidelberg
2. Hodges B (2008) Sea water farms Eritrea 1998–2003. Atlantic Greenfields
3. Food and Agricultural Organization of the United Nations (2003) Review of world water resources by country, Rome, 97 p
4. Lenntech (2011) Use of water in food and agriculture 5 p. http://www.lenntech.com/water-food-agriculture.htm
5. World Resources Institute (2008) World resources 2008—root of resilience—growing the wealth of the poor. In: collaboration with the United Nations Development Programme, the United Nations Environmental Programme, and the World Bank, Washington, D.C., pp 210–213
6. Population Reference Bureau (2012) 2011 World population data sheet: population, health, and environment, data and estimates for the countries and regions of the world. Washington, D.C.
7. Gleick PH (2012) The world's water, 2012 vol 7: the biennial report on fresh water resources. Island Press, Washington, D.C., 277 p
8. Blachman A (2008) How to make water from thin air. cleantechnica.com/11/05/how-to-make-water-from-thin-air
9. Waldoka EZ (2009) In world first, Israeli know-how helps Indian village get water from air, Jerusalem post. http://www.JPost.com/servelet/Satellite?cid=1233304849721&pagenames=JPost%2FJPArticle%2FShowfull
10. UNESCO (2012) Managing water under high uncertainty and risk, 4 edn, vol 3. World water development report, Paris, 866 p
11. Bruinsma J (2009) The resource outlook to 2050. By how much do land, water and crop yields need to increase by 2050. Expert meeting on how to feed the world in 2050. FAO of the UN, Economic and Social Development Department, Rome, 33 p
12. Food and Agricultural Organization of the United Nations (2006) Livestock's long shadow—environmental issues and options. Rome, 390 p. ftp.fao.org/docrep/fao/010/a0701e/a701e.pdf. Accessed 14 Mar 2014
13. U.S. Department of Agriculture/Foreign Agricultural Service (2013) Livestock and poultry: world markets and trade. Washington, D.C., 29 p
14. Foer JS (2009) Eating animals. Little Brown and Company, New York, p 329 p
15. Gietema B (2005) The basics of chicken farming (in the tropics). Agromisa Foundation, Wageningen, Netherlands, 170 p. http://www.ruralfinance.org/fileadmin/templates/rflc/documents/
16. Gietema B (2006) The farm as a commercial enterprise. Agromisa Foundation, Wageningen, Netherlands, 99 p. http://www.fastonline.org/CD3WD_40/LSTOCK/001/agrodocks/EM-32-e-2005-digitaal.pdf
17. Patel P (2010) Solar-powered desalination. Technology review, MIT, 2 p
18. Lee E (2010) Saudi Arabia and desalination. Harvard Int'l Review, 23 Dec

19. Roberts DA, Johnston EL, Knott NA (2011) Impacts of desalination plant discharges on the marine environment: a critical review of published studies. Water Res 44(18):5118–5128
20. National Research Council (2008) Desalination: a national perspective. National Academy of Science Press, Washington D.C, p 290 p
21. Cyna B, Chagneau G, Bablon G, Tanghe N (2002) Two years of nanofiltration at Mery-sur-Oise plant, France. Desalination 147:69–75
22. Nicoll H (2011) Nanofiltration membranes. Water Wastes Digest 16:38
23. Pigeot J (2000) Le Figaro, June 20, p 14

Chapter 3
Strategies to Increase Food Supplies for Rapidly Growing Populations: Crops, Livestock, and Fisheries

3.1 Introduction

The Food and Agriculture Organization (FAO) of the United Nations estimates that by 2050, food production will surpass 2009 levels by 70 % globally and by 100 % in developing countries. This may supply the needs of a growing world population that is projected to be more than a third larger (2.6 billion more people) by 2050 than in 2013. It will likely ease the food security risk for Africa and Asia, regions where populations will grow most and where the majority of the world's one billion malnourished population live today (e.g., Asia with 578 million and sub-Saharan Africa with 239 million). To achieve this goal means that cereals production will have to increase by one billion tonnes and livestock products by 200 million tonnes, annually [1]. This will have to be done as competition for land and water intensifies for other uses (e.g., feedstock, biofuels). Competition will also come from aquaculture that has grown 6.6 % annually during 1970–2008 and is expected to continue increasing its contribution to the global food fish supply. Further, rising incomes and changing dietary tastes in developing nations (e.g., inclusion of more beef and dairy products) may temper the easing of food security risks so that even with a doubling of agricultural production by 2050, malnourished persons will still number 370 million citizens. This is just under 4 % of the projected 2050 global population. Although not acceptable, this is a notable improvement over the 14 % malnourished persons on earth in 2013.

During the past 50 years, food production has grown by two and a half to three times, but cultivated land in this period grew by only 12 %. The FAO estimates that more than 80 % of the added yield is from improved productivity of current cultivated land. About 40 % of this advance in food production was from irrigated land. The rate of food production of about 3 % annually was greater than that for the global population, but in recent years, this rate has dropped to about half. Population expansion now exceeds growth in food production [1]. Clearly, investment has to be made to impart practical and technological land and water management

© The Author(s) 2015
F.R. Siegel, *Countering 21st Century Social-Environmental Threats to Growing Global Populations*, SpringerBriefs in Environmental Science, DOI 10.1007/978-3-319-09686-5_3

expertise to less developed and developing countries that need it most so that they can strive to increase their rate of agricultural output to match or exceed the expansion of their populations. Economic analysis puts this investment at almost US$1 trillion for irrigation development and management plus another US$160 billion for land protection and development, soil conservation, and flood control. These funds are expected to originate from governments, institutions such as the World Bank and regional development banks, international aid organizations, and multinational food-producing companies. The chapter sections that follow will consider corrective solutions to the problem of producing enough food and the capability and freedom to transport foodstuffs to where they are needed in order to provide for all of today's citizens as well as those that will comprise future generations.

3.2 Strategies to Increase Crops

3.2.1 Increase the Acreage of Arable Land that Can Be Tilled

An obvious strategy to increase food supplies is to increase the area of arable land that can be farmed. Arable land as defined by the FAO is land under temporary agricultural crops, as land with meadows for pasture and mowing, as land under market or kitchen gardens, and as land fallow for less than 5 years. Crops are classified as temporary or permanent. Permanent ones are long-term crops that are planted once and that do not have to be replanted for several years. These are mainly trees (e.g., oranges, apples, cocoa, palms [dates], bananas, coffee), but also include crops from shrubs and bushes, vines, stemless plants, and flowers (e.g., blueberries, blackberries, grapes, pineapples). Temporary crops are planted and harvested each year to maintain production (e.g., maize [corn], rice, wheat, soy).

Of the approximately 13.9 billion hectares (ha) of land on earth, 11 % or about 1.53 billion ha is under cultivation. Of these, 1.23 billion ha is rain-fed and 0.3 billion ha is in irrigation. In recent years, the area of rain-fed cultivation has dropped slightly, whereas the area of irrigated cultivation has increased more than 100 %. Indeed, more than 40 % of the increase in food production in the past half century is from irrigated land that has doubled in area. There are large zones of potentially useable agricultural land on earth, but only small areas of these can be used to expand cropping. This is because much of the earth's land area is forested and has to be preserved, whereas other areas that are arable are protected because of environmental concerns (and laws), and still other arable areas are occupied by urban centers or set aside for urban development. About 90 % of the additional land that could be put in cultivation (e.g., because of water availability, temperature, and climate factors) is in sub-Saharan Africa and Latin America. Half of this is in seven countries (Brazil, the Democratic Republic of the Congo, Angola, Sudan, Argentina, Columbia, and Bolivia). The countries in northeast Africa, Southeast Asia, and Southwest Asia have little land that can be put in cultivation [2].

Although the world produces enough food to sustain a population of more than 7 billion people, the production will have to increase by 60 % above the 2005–2007 level (given as 70 % above 2009 levels [2]) to feed a population that is predicted to increase to more than 9.7 billion persons by 2050 and could reach 10.3 billion or more by 2100. Production of enough food to feed the 2012/2013 population does not solve the hunger problem. As already noted, the world has close to a billion people who suffer from chronic malnutrition. Unhindered distribution of food can help alleviate the food deficiency problem. That we will be able to increase global food production 60–70 % by 2050 does not mean that the world's future hunger problems will be solved. Political, economic, and logistics issues will have to be resolved in order to ensure that food reaches the people suffering or close to suffering malnutrition.

3.2.2 Increase Crop Yield and Nutritional Value: Improve Seed that Protect Crops from Harm by Biological, Physical, and/or Chemical Agents

Biologists are continually working to improve crop resistance to elements (e.g., weeds, pests, extreme weather) that reduce the yield and nutritional value of foods that are staples for billions of people worldwide such as maize [corn], rice, wheat, soy, and other grains. This is being done through hybridization. Hybridization is the process of combining two complimentary DNA or RNA molecules and allowing them to form a single double-stranded molecule through base pairing. Simply stated, hybridization is the process of interbreeding between genetically divergent individuals from the same species (conventional—intraspecific hybridization) or between individuals of different species (genetically modified—interspecific hybridization). There is no assurance of the results of hybridization. The offspring may be fertile, partially fertile, or sterile. Biologists use traditional hybridization, marker-assisted selection (MAS) for traditional hybridization, or genetic engineering/manipulation. They work to produce seeds that give higher yields of crops with higher nutritional values and seeds that impart properties to plants that make crops resistant to disease, insects and other pests, weeds, drought, submersion from floods, and salt and cold tolerance. The ability to protect crops from one or more than one of these biotic or abiotic stresses will, by itself, enhance crop yield and nutritional value. The hybridization processes discussed in the following paragraphs will have to continue to breed plants that will be resistant to and tolerant of new crop pests and diseases and other changes in growth environments that evolve as a result of global warming and consequent climate change.

3.2.2.1 Traditional Hybridization

Traditional (conventional) hybridization is a long-term process. It may take many years to breed a plant with a specific characteristic that will be a successful fertile parent from which to recover a new seed stock. It involves discovering a plant from the same species that carries a resistance or tolerance trait that protects the plant from sickness, damage, or destruction. It then requires the breeding of the trait(s) into a cultivar that protects a crop from harm, thus preserving and increasing the food stock. The breeding is done via controlled pollination using natural cross-pollination by insects (e.g., bees, wasps), by wind (e.g., for maize/corn), and by self-pollination (e.g., for wheat and rice). Experience is a guide as to how best to synchronize flowering and pollination techniques. When the trait has proven transferable and the seed that carries it is fertile, the hybridized intraspecies will be the parent for an improved seed stock. The expectation is that this will increase food access for today's populations and help provide food for growing world populations. There have been numerous advances in many less developed and developing countries for crops that have better yield and nutritional values and other desired characteristics and that are staples for billions of consumers worldwide. In addition to controlled pollination, scientists have worked with chemicals and radiation to induce a mutation (mutagenesis) that will carry a desirable property but does not use genetic engineering gene splicing. As emphasized earlier, the traditional hybridization process is slow. In recent years, the slowness of this process has been overcome to a good degree by MAS, a method that speeds up the conventional procedure to improve commercial cultivars.

3.2.2.2 Marker-Assisted Selection for Traditional Hybridization

DNA transmits genetic information. The DNA is in chromosomes that are in the nucleus of every cell in an organism. It contains all of the chromosomes that comprise the genome of the organism in question. MAS is a method in which a derived molecule marker based on DNA variation is used for the indirect identification and subsequent selection of a genetic determinant (or more than one) that makes a crop species resistant to harmful organisms or conditions. MAS speeds up traditional hybridization by using DNA from specimens at the seedling stage to identify markers that carry the sought for crop traits. This cuts in half the time to find individuals in cultivar species or wild species that have the genetic markers that can be used to improve crop characteristics and that will help increase food supplies [3]. Those intraspecies selected individuals are crossbred with commercial varieties through MAS hybridization to develop the next generation of cultivated crops. The same species with improved traits will be parents for a future generation of food crop seed stocks. A drawback to the use of MAS by developing nations is the initial investment and technical sophistication necessary to extract the genome of the target cultivar. This process can be affected by environmental parameters and expected progeny differences.

Successes of MAS hybridization include the development of a pearl millet hybrid with resistance to downy mildew disease in India [4], new rice varieties with resistance to bacterial blight in India [5], and rice with submergence tolerance in the Philippines [6]. The application of MAS hybridization to cereals is important to increase their supply [7]. The method is being used to transfer sought for characteristics to perennial, subsistence, and cash crops in developing countries. These crops include tropical fruit trees (e.g., bananas, mangoes), and coffee, tea, cocoa, coconut, and oil palm [4]. In 2012, researchers reported that they identified a gene that enhances root growth and access to more phosphorus that can be absorbed from soils with low phosphorus content by analyzing part of the Kasalath rice species DNA [8]. This gene is absent from the rice reference genome. They named it phosphorus-starvation tolerance 1 (PSTOL1). This gene also gives the rice plant tolerance to winter conditions and accelerates maturity. Using traditional cross-pollination hybridization, the scientists introduced the gene into some rice types in Indonesia, the Philippines, and Japan with the result that yields increased by 20 %. This genetic modification does not represent genetic engineering that splices a gene into a plant's DNA, thus avoiding the GMO controversy in many countries.

3.2.2.3 Genetic Engineering: Genetic Manipulation/Modification

The life, growth, and unique characteristics of organisms depend on their DNA. Genes are segments of DNA that are associated with specific traits or functions. Many can be identified once the genome of an organism has been worked out. The process of genetic engineering involves genetic manipulation or modification (GM) in which a gene that carries a desirable property in one species is cut from it and spliced into the DNA of another species that lacks the property. For life forms, specifically food crops, this interspecies manipulation creates a genetically modified organism (GMO) that will be the parent for an improved seed stock. Researchers involved in genetic engineering at biotech companies such as Monsanto, Bayer, Dupont/Pioneer, Dow/Mycogen, and Syngenta believe that this is an efficient and inexpensive way for agriculture to increase the global food supply by creating seeds with one or more of the attributes that enhance the yield and nutritional quality of a food crop. These companies assume intellectual property rights to crop seeds they engineer that have unique traits that improve a crop's success. As listed in the paragraph on traditional hybridization, these characteristics are better nutrition, resistance to disease, insects and pests, weeds, tolerant of herbicides, drought, temporary submersion during flooding, and salt and cold tolerance. GM research also includes a pharmaceutical path that would produce edible vaccines in tomatoes and potatoes and a phytoremediation path that would increase the capacity of plants to absorb toxins such as those known to take up heavy metals from soils [9].

Genetic engineering has focused on improving seed quality for basic staples such as maize [corn], rice, soy, and wheat for populations in less developed and developing nations in Asia, Africa, and Latin America. Genetic manipulation has successfully produced seed for disease-resistant Asian rice, [10] for Asian rice with higher levels of nutrients, [11] but was unsuccessful in producing seed for a sweet potato variety that resists a common African virus with a successful seed produced by traditional crossbreeding [12]. Other genetically manipulated commodities with different beneficial properties that improve the quantity and quality of crops in addition to those already mentioned include alfalfa, Hawaiian papaya, sugarcane, sugar beet, tomato, potato, zucchini, and sweet peppers [13]. Countries such as Brazil, China, and India have supported intensified research on genetic engineering of indigenous crop species. In this way, they can focus on improving traits of crops in their countries and be independent of intellectual property rights (patented transgenic forms) of the biotech companies and the economic payout for company seeds. Brazil bred a herbicide-resistant soybean, China developed a GM rice with higher yields, greater productivity, and need for less fertilizer and pesticide, and India produced a rice that is resistant to disease. In light of these advances, Monsanto donated a drought-resistant maize it produced for Africa in what could be considered a marketing effort to allow it more entry into African countries.

There have been great successes in the 15-year transgenic crops that have been planted. Indeed, three-quarters of the world's soybean crops are a genetically modified variant as is one-half of the global cotton crop and one-quarter of the world's maize/corn crop. In the USA for 2012, 93 % of the planted soybean was HT (herbicide-tolerant) soybean, 82 % of the cotton planted was HT cotton and 75 % BT (bacteria-tolerant) cotton, and 76 % of the corn planted was BT corn and 85 % HT corn [14]. Transgenic crops are grown in only 25 countries of the more than 200 on earth. Many countries, such as those of the European Union and a great majority in Africa, are against planting these crops and against allowing sale of food products manufactured with them. The major problem that these countries have with GM crops is the risk to non-GM natural crops by pollination from GM crops by wind-carried pollen, by insects carrying pollen, and by birds carrying pollen and how this cross-pollination will affect natural crops [15]. There is an additional risk of wind-borne pollen being carried to villages near farms. Such an occurrence may have been the cause of sickness and death in humans and in farm animals in the Philippines where the maize seed Dekalb818YG with Cry!Ab from the soil bacteria *Bacillus thuringiensis* had been planted [16]. Can transgenes be isolated? How? Can they be contained? How? GM maize grown as feed was used to prepare foodstuffs sold by a national chain in the USA. People who ate food with the GM maize got sick, and both the people and the chain were compensated by the grower/seller of the maize. There was a recall of the products made with the GM maize. Can the GM crops be safely handled? Time will tell as we gain knowledge of the effects that GM foods may have on life forms and environments [17].

3.2.2.4 Vertical Farming

In his 2010 book "The Vertical Farm: Feeding The World In The twenty first century," Professor Dickson Despommier of Columbia University champions the concept of vertical farming in especially engineered skyscrapers in urban centers

[18]. This, he believes, can boost food security for their growing populations estimated to reach 70 % of the global population in 2050 or about 6.8 billion people. The plant growth would be in a controlled environment with temperature, humidity, nutrient delivery, and airflow keyed to crops grown on different floors to maximize crop yield and quality throughout the year. In this scenario, there would be minimum use of chemicals to kill weeds, plant fungi, and pests that may show up, and water and nutrients would be recycled. Plants would be safe from extreme weather events such as drought, cold snaps, hail, and high winds. Electricity to power the system could be supplied in part by solar energy installations on sky-scraper rooftops. In addition, transport costs to move food from source to consumer would be reduced. A major problem in addition to initial capital investment in a vertical farming project is that there are no insects to pollinate crops so that mechanical or manual pollination would be necessary. Pilot projects are ongoing in Singapore and Sweden.

3.3 Agricultural Practices to Prevent Soil Erosion and Maintain Soil Fertility Managing Water and Farming Methods that Protect Soil and Nearby Ecosystems

3.3.1 Soil Preparation for Planting, Cultivating, and Harvesting Crops

There are two principle soil preparation techniques: tilling and no-tilling (also called tillage and no-tillage). Each one has its benefits and drawbacks and is selected for use according to soil and climate parameters and economic considerations.

3.3.1.1 Tilling

Tilling prepares a soil for farming by digging into and overturning it, thus breaking it up into a rough surface. This exposes more surface area of the soil to weather and makes it more susceptible to erosion by rain runoff and wind. On the positive side, tilling also aerates and loosens soil to make sowing seed easier. Tilling also tears up weeds that would otherwise drain nutrients from a soil and crop residue and mixes them into the soil. The residue represents at least 30 % of the organic matter cover (~1,100 kg/ha or 1,000 lb/acre). This organic matter degrades and releases nutrients, thereby improving soil fertility. Tilling exposes disrupted soil to the sun that causes a loss of soil moisture to evaporation, a loss of nitrogen and carbon nutrient compounds, and a loss of soil biodiversity. Tilling facilitates a high amount of agricultural chemical runoff. Multiple passes with tiller, planter, cultivator, and harvesting machinery compact a soil. Irrigation equipment may do the same. Soil compaction may make it difficult for plants to root deeply enough to take up nutrients and water sufficient to support optimal growth. Obviously, it is necessary to know the climatological conditions of the farming region (e.g., precipitation, hours of sunlight, temperature during day and night, length of growing season) and the types and properties of soils that are being farmed. This allows farmers to pair a crop with a soil. The cost of diesel fuel for the farm equipment that accomplishes the tasks just mentioned increases the costs of agricultural products.

3.3.1.2 No-Tilling

No-tilling is a practice that sows seeds by using slice and seed or puncture and seed mechanized equipment that really does not dig up a soil, thus reducing erosion possibility by runoff or wind. Poking a hole in the soil and planting a seed in it has been used for centuries by indigenous groups worldwide and is still used today. There is an increase of water and decaying crop residue (nutrient source) in no-till soils. Residue from recent harvests is left at the surface and serves to trap and hold rainwater and upon decomposition adds to the soil nutrient contents and its fertility. There is also an increase of biodiversity in a no-till soil, and this also aids soil fertility. The single pass of the planting equipment reduces the cost of diesel fuel for the agriculturalist. Because the soil can be packed down, it is often necessary to plant crop varieties that tolerate such soil. A plant variety that is selected for no-till farming will be determined by soil type and climate conditions as noted in the previous paragraph. However, a major drawback to no-till agriculture is that it does not take care of the weed problem so that heavy use of herbicides is necessary. Globally, there has been a significant increase in no-till agriculture. There was a 74-fold increase in no-till agriculture in Latin America that increased from 670,000 ha in 1987 to 49.6 million ha in 2008. For the same interim, the US acreage under no-till farming increased 6.5-fold [19]. Worldwide, there were 105 million ha under no-till agriculture in 2008. An excellent evaluation of no-tillage crop production was published in 2007 [20].

3.3.1.3 Multispecies or Poly-varietal Cropping

A good agricultural practice that can be economically beneficial is multispecies cropping. Mono-cropping presents a risk of total loss if a disease attacks the plants. Mono-cropping was a bane to sweet potato production in Africa until, as noted previously, a sweet potato plant was crossbred to resist a common African virus [12] and was successfully planted, cultivated, and harvested. Multispecies cropping has the benefit of yielding a crop even when one crop in the same agricultural area fails.

Intercropping is another good way to increase crop production and protection on farmland. Intercropping means that agriculturalists grow two different crops with different requirements on the same acreage. One method would be to grow shallow-rooted plants together with those having deep roots so that crops do not compete with each other for soil nutrients. This also avoids a problem that could arise from mono-cropping.

3.3.1.4 Intensified Agriculture

Intensified cropping is designed to grow crops that give higher yields on the same amount of acreage but crops not necessarily of high quality. Intensive agriculture is often located close to consumer markets. Because it costs less to grow crops and transport them to sale locations, the cost to the consumer is less. Intensive agriculture has benefits and drawbacks as a good agricultural practice. It requires a high capital investment and a lot of labor. It is mechanized and uses high-efficiency machinery for planting, cultivating, and harvesting, and, where necessary, efficient irrigation equipment to manage a water supply. The acreage is fallow for less time, and there is a heavy use of agricultural chemicals (pesticides, insecticides, herbicides, fungicides, and chemical fertilizer). This is a negative for ecosystems. For example, pesticides kill off beneficial insects as well the target ones and in general decrease biodiversity. Also, as already emphasized, agricultural chemicals can run off into streams, rivers, and lakes and can seep through soils with surface waters to invade groundwater. Many intensive farm fields are close to towns and cities so that depending on wind conditions, pesticides and other chemicals can be carried to population centers and harm people that inhale them. The workers that apply the pest-killing chemicals are at risk if protective ventilation masks are not used. There is an unseen pesticide residue on the crops that does not wash off easily and that can, over time, harm consumers that regularly eat the crops so that thorough washing of produce is essential. When there is a high use of irrigation, salination can coat roots with salts over time and reduce crop productivity. Intensive agriculture is useful but only if properly managed. If this is not done, the soil ecosystem can fail and not support further cropping. Soil erosion control should be exercised, and, if feasible, organic fertilizer and organic pesticides should be used on crops. If organic chemicals are not used, only the minimum amounts of agricultural chemicals should be applied to maintain optimal growth and thereby reduce any runoff or seepage through soil and thus protect surface water and groundwater.

3.3.1.5 Sustaining Soil Fertility

Nutrients including carbon, nitrogen, and phosphorus are taken up from the soil during plant growth together with trace metals (e.g., potassium, calcium, magnesium, sodium) and heavy metals that are essential to a plant's health. Nutrient that is extracted from soils by plants as they grow must be replaced to sustain a productive soil system. This is done in two ways. First is the use of man-made chemical fertilizers. The amounts that are applied to a field, and how often they are applied, are dependent on the concentration already in the soil, the crop that is being sown, the climate, and other factors. This can be determined by county agents or other government experts who are knowledgeable about a region's agricultural characteristics. The minimum amount of man-made fertilizer used to stimulate optimum crop growth will benefit a farmer by cutting expenses for fertilizer. Second is the use of organic fertilizer that is composed of crop residue, compost, and dried, treated manure that is mixed into a soil. As before, a county agent can measure the nutrient level in a soil and determine whether a supplement to an organic fertilizer is necessary given the crop a farmer plans to grow in the soil.

Biodiversity in a soil adds to a soil's capacity to support plant growth. Use and overuse of man-made chemicals to control weeds and insects can severely harm soil ecosystem biodiversity. Thus, as noted before, where possible if organic methods are not practical, it is preferable that agriculturalists apply herbicides or pesticides to crops only in amounts that will protect a crop from such problems and help sustain soil ecosystem biodiversity and fertility.

3.3.1.6 Optimizing Water Use

Efficient water use is good farming practice. For rain-fed agriculture that is known to have periods of shortfall, it is prudent to have a reservoir where excess rainfall can be directed and from which water can be supplied as needed during the shortfall periods. Arm irrigation can be done earlier in a day to minimize water lost to evaporation. Water can also be loaded into soil via arm irrigation during non-growing seasons. Drip irrigation that delivers the correct amount of water at the base of a plant to keep it healthy is the most efficient way to manage an irrigation water supply. There is a capital investment to set up a drip irrigation system and maintenance costs to keep it operating efficiently. However, it is economically beneficial to agriculturists who recover their investment by not having to pay for water lost to evaporation or to runoff that does not nourish a crop.

3.3.1.7 Managing the Problem of Soil Salination

Salination or precipitation of solutes from irrigation waters, from rainwater that dissolves salts as it seeps through a soil to nourish crops, or from rising water table and capillary action causes crusting on plant roots. This problem affects large areas worldwide. For example, more than 50 % of the soils in Australia, Africa, Latin America, and the Middle East suffer from salination. Over time, as the solute precipitation continues, the salt encasing the roots thickens to the degree that nutrients can not be taken up from an otherwise fertile soil so that crop yield and quality fall and plants are stunted. Therefore, monitoring the salt condition of a soil is important in order to be able to remediate a salination threat to a crop before it becomes a reality. The salt can be removed by regularly scheduled flushing to desalt a root zone so that nutrient uptake by a plant is not limited or stopped [21].

Salination can be managed, and crop productivity sustained by the installation of a subsurface drainage tile network. When a soil is flushed, the network moves irrigation and/or percolated rainwater out of soils. This reduces the possibility of short-term salt build up on crop root systems. A large capital investment is necessary, but the cost/benefit numbers favor the investment in not too long a term even when a 10–20 % annual maintenance cost is factored in. Conditions such as soil type, climate, and crops grown determine the design for the depth of a tile network and tile spacing. In projects funded by the World Bank for irrigation-based agricultural development, the costs for the installation of drainage tile networks ranged from US$1,100 per hectare in Pakistan, to US$1,900 per hectare in Peru, and between US$1,800 and US$2,300 per hectare in Egypt for a project covering millions of hectares. Keeping plant root systems free of the salination problem keeps agricultural fields productive. Application of other crop-enhancing methods as cited in previous paragraphs should lead to increased productivity and add to the global food stocks that will be necessary to feed growing global populations.

3.3.1.8 Minimizing Pre-distribution Food Loss

The amount of food lost each year at farm or depot storage facilities and at distribution points where the food has been sent for transport to people that need it now, and those that will need it as global populations expand, is a huge waste. The spoilage of food is from rodents, insects, bacteria, mold, moisture, and heat. For example, rats eat or contaminate 29 % of food (grains) in storage facilities. Investment in storage facilities with tight construction to limit access by animals, and with air-conditioning and dehumidifier capability to prevent loss from heat and moisture, will save large amounts of food worldwide. The problem that has to be overcome to achieve this is not the construction of storage facilities alone but one of electrification. Electricity that is necessary to power cooling and dehumidifier units is not often available at storage locations in many food-recipient countries. If access for rodents is halted or minimized and the electricity problem solved, this will increase the world's available food supply significantly and should reduce global malnutrition.

Food does not always arrive where it is needed, adding to the problem of supply. There is skimming by corrupt personnel at storage facilities for subsequent sale on the black market. In other cases, there is a commandeering of comestibles in transport by government/clan/tribe/pirate edict or force of arms. Again, those who need the food supply most do not receive it because it is a commodity that can be sold by bandits. If governments purportedly value and follow the rule of law and have the will to do so, they can assure that food supplies sent from other regions in their countries or from other countries to feed their hungry and malnourished citizens will arrive safely and intact at their designated destinations.

3.3.2 Increasing Livestock (Protein) Contribution to the Global Food Supply

Although the rate of population growth is declining, populations continue to grow in most of the world. In 2009, FAO reported that this, together with rising affluence and demographic movement to urban centers, especially in Asia, has increased the demand for livestock meats. This is a sign of economic well-being for some and as a source of protein and essential trace metals for better nutrition for others (e.g., there is more iron in meat, milk, and eggs than in plant-based foods). The main sources for the meat are beef cattle, poultry, pigs, and sheep. Cattle, and to a lesser degree sheep, also provide milk and use it to produce protein-rich dairy products. Chickens provide protein-rich eggs. At this time, suppliers are able to provide this food supply although it is not available to all because of location, infrastructure failings, and non-affordable costs. As a result, and as reported before, one person in seven on our planet, especially babies and small children, suffers from chronic malnutrition caused by a lack of enough food that is nutritious. Given the global population that is projected to reach 8.6 billion people in 21 years (2035), at least 9.7 billion by 2050, and more than 10 billion by the end of the twenty-first century, the concern is whether livestock production can match or exceed population growth. In addition, there is a question of whether this can be done so as to make meats more affordable to citizens with limited economic means. Finally, if livestock production increases to be able to feed the world's future populations, what will the environmental costs be and can these be mitigated by controlled management of the livestock sector?

There is no doubt that meat production can increase to provide protein-rich food that also contains essential macro- and micronutrients. Global meat production (from cattle, sheep, pigs) remains level over the past two decades except for growth in Chinese pig output. This may be a result of human health concerns (e.g., cholesterol levels), or of sociocultural beliefs (vegetarianism, religion), or caused in part by animal welfare groups and environmental legislation. Nonetheless, meat production is in a recovery stage with some rise in production expected for 2012–2013. Conversely, poultry meat production (mainly chicken) has increased at a rate three times that of population growth during the past 50 years and during 2009 rose 5.8 % [2]. Whether total meat production can increase to satisfy the needs of the earth's citizenry at affordable costs in 2035, 2050, and beyond remains a question. There are basically two paths to increase meat productions, one physical/biological and another biological/pharmaceutical. Both have drawbacks.

3.3.2.1 Physical/Biological

Intensive, industrial-size animal husbandry operations can promote numbers of stock and hence the meat supply. This is being done successfully in many developing countries by moving animals from grazing and yard-kept growth sites to densely populated pens where they feed on nutrition-rich cereals and high-protein oil meal (soy). This has caused a competition between farmland use to grow crops for human consumption and that used to grow feed crops/cereals and oil seeds for meat production. Another competitor for farmland is the growing of plants as a source of biofuels, a factor that also threatens food security and affordability. During 2013, the USA cut the subsidies for biofuel cropping so that much farmland will revert to growing food for people. The drawbacks here are the necessity to use or create more agricultural land in addition to the 40 % used for animal husbandry. This has meant deforestation in some Latin America and Caribbean nations. Overgrazing and harm to the land and soil security is a problem where animals are not penned up but are kept in large tight herds as they feed.

3.3.2.2 Biological/Pharmaceutical

Meat production increases as growers become more efficient. Many intensive animal farming projects employ advances in breeding, genetic gains, and nutrition, to have a better selection of animal (cattle, chicken, pigs, sheep) to grow for meat production. To keep the animals healthy in the close quarters where they are penned up and so that they grow faster, they are fed antibiotics added to their feed and water. The drawback here begins with the fact that food animals consume about 80 % of antibiotics sold in the USA. Two-thirds of these drugs are identical or similar to drugs prescribed by doctors or used in hospitals. The red flag here is that even a low, short-term dose of in-feed antibiotics increases the abundance and biodiversity of antibiotic-resistant genes, including resistance to antibiotics that were not administered, as well as increasing the abundance of *Escherichia coli*, a human pathogen [22]. An earlier study found that soil bacteria surrounding conventional pig farms and downstream waters carry 100–1,000 times more resistant genes than do the same bacteria around organic pig farms [23]. These studies concluded that antibiotics used in animal farming contribute to the resistance of bacteria (superbugs) to drugs. The more that antibiotics are used, the less effective they become. The resistant bacteria can access citizens through food or from the disposal of excretions that carry the bacteria to soils and waters. In the USA, 23,000 people die annually from drug-resistant bacteria, many from hospital-acquired infections. The conundrum is that meat production would fall precipitously if antibiotics were not used. The European Union banned the use of antibiotics on healthy farm animals in 2006. The US Federal Drug Administration (FDA) does not want to make many antibiotic drugs less useful for human health needs. The FDA requested that congress allows edicts that stop the use of antibiotics to help cattle feed well and grow faster and that animal farmers use the drugs judiciously (not in excess) only to keep the animals disease free and then under the supervision of a veterinarian. Industry interests and political influence are working against this even if it is clearly a threat to public health. Whether congress votes to protect public health or defers to political influence (e.g., donations for reelection) is still in discussion. Should other countries go the way of the European Union and protect the health of their citizens?

Many are assessing the long-term viability of their very intensive farming techniques in order to make changes that will sustain environmental vitality. For example, because of the need to keep their water supplies useable, effluents are being collected, treated, and recycled into the system (see Chap. 2). Feces can be collected, treated, dried, and used as fertilizer or as briquetted fuel in combustion furnaces. In this way, the intensive farming industry is working to protect water, soil, air, and other natural resources from pollution that could ultimately negate increases in meat production.

There are about 1.5 billion cattle worldwide. They emit an estimated 57 million tons of methane to the atmosphere annually through burping and flatulence. This result of bovine enteric fermentation may be responsible for 72 % of all methane emitted to the atmosphere. Methane is a greenhouse gas 20 times more powerful than carbon dioxide in abetting global warming. Professor Stephen Moore, formerly of the University of Alberta, Canada, and now at the University of Queensland, Australia, and his team used traditional techniques to breed cows that produce 24–28 % less methane than cows not bred to do so. This means about 15 million tons less of methane would be emitted to the atmosphere if the researchers' concepts prove successful and are applied in the global cattle industry. The scientific team examined genes responsible for fermentation and methane produced by a cow's four stomachs to find animals with the diagnostic marker for the breeding. Their studies also reported that when cows have a diet rich in energy grains and edible oils, there is less fermentation than from cows that feed on grass or lesser-quality fodder [24]. A few dairy farms in New Hampshire, USA, reduced cows' methane emissions by 12 % by using a feed of alfalfa, flax, and hemp. Other ruminants (e.g., pigs, sheep, goats) add to the methane contribution to the atmosphere and warrant similar research as done on cows. A significant decrease in greenhouse gas emissions to the atmosphere can help slow the rate of global warming, ultimately to stabilize it, and allow for improved planning for the earth's inhabitants' long-term future.

3.3.3 Fishery Contributions to Global Food Stocks Can Help Feed Growing Populations

An estimated 2.9 billion people, mainly in Asia, get about 20 % of their protein intake from food fish [2]. This is one of the regions that will have a large growth in population as we approach mid-century. The amount of food fish in our food supply is increasing, but much of the increase is not from capture fisheries but rather from marine and on land aquaculture farming. By 2009, aquaculture contributed 38 % to the fish supply (56 million tonnes). Nearly half of this represented the fish supplied for peoples' food, while the rest was for fish meal, for fish oil, and for animal feed (e.g., for chicken, pigs, salmon, and shrimp). In 2009, FAO estimated that 122 million tonnes of food fish were captured, mainly in oceans (85–95 million tonnes) with significant contributions from estuaries, lakes, and rivers [25]. Most of the growth in aquaculture production during 1990–2009 was in developing countries. This was the result of improved breeding techniques in the growth ecosystem itself and in the feeding phase. The growing Asian population accounts for 90 % of the aquaculture production worldwide with China and its 1.35 billion person population responsible for two-thirds of this. Inland capture fisheries contributed 10 million tonnes that was less than 10 % of the world's capture tonnage. About one-half of the marine and inland fish catch was from small fisheries.

Capture fisheries growth has not kept pace with population increase and is not likely to do so. This is because of more stringent and enforced international regulations, scarcity of target fish because of overfishing, and rising costs for developed countries. Conversely, the growth in aquaculture production has been greater than the rate of population expansion. Carp farming represents 60 % of the world production as well as the major share of high-value fish production (e.g., salmon, shrimp, and scallops).

The state of major ocean fish stocks tells the story of why an increase in ocean capture is not likely in the near future. Thirty-two percent of the ocean fisheries were overexploited, a threefold increase from the early 1970s. This depleted the fisheries stock. Fifty percent was fully exploited. This leaves and only 15 % of the ocean fish stock as underexploited to moderately exploited in 2008 against 40 % in the mid-1970s [25]. Overfishing has been the problem that is being dealt with through international agreements. There are positive advances in rebuilding marine stocks that have been overfished.

One impediment to restoring depleted fish stocks at a reasonable pace is illegal, unregulated, and unreported fishing. Fisheries experts judge that 12–22 % of fish trade is through this source [25]. To try to reduce this problem, countries have passed laws, while others are being coerced to pass laws that regulate fishing lest they be frozen out from international trade in fish or fish products. This presents another problem because many countries do not yet have the capacity to surveil and thus monitor and use their naval or coastal police forces to control illegal fishing. Certainly, the ability to source fish and fish products from legal fisheries can identify those that originate from illegal sources. Developed countries have to assist developing countries to improve their capabilities to discourage illegal fishing.

It is probable, even likely, that the amount of fish in the global food supply can increase to help feed the projected world population of 9.7 billion people in 2050. This, if aquaculture continues its rate of growth and if marine fish stocks are allowed to rebuild to levels that allow quota controlled harvesting that sustains steady productivity.

Overfishing decreased stocks of high-value and low-value food fish and in some cases disrupted the marine food chain so that some predators had to find new prey to survive. Only, reduced hunting quotas or a ban on quotas of certain food fish and time to let them restock can help sustain this food supply. If this is not mandated and enforced, some reports in the 1990s estimated that by 2025, 76 % of our global fisheries could be lost. Given the enforcement of international (and national) regulations, fisheries scientists believe that fish stock rebuilding can take place using individually tailored management of the process. Researchers report that marine fish stocks vary from being stable, or declining or collapsing because of the lack of harvesting control, with some in a rebuilding mode because of following of harvest quotas and other managerial practices. They document fish stock rebuilding for 1 % of the marine fisheries and report that recovery is taking place in other fisheries that were in decline or nearing collapse because rates of catch recently have lessened to rates that are below those necessary to allow restocking and future sustainable yield. They estimate that 63 % of the fisheries assessed to date require rebuilding [26]. Rebuilding the fish stocks can be done but takes time to assure that their future harvests can be sustained. Management practices to achieve this end include areas closed to fishing and restrictions on the sizes and characteristics of gear used (allows for selectivity of the species captured). To maintain a fishing community with a liveable income, there can be a sharing of the allowable quotas catch among the fishermen until a stock of targeted food fish is rebuilt and all can have their fixed quotas of sustainable fish populations.

In a 2012 study, another group of fisheries scientists reported that many well-assessed ocean fisheries are moving toward sustainability [27]. However, these yield less than 20 % of the global catch. The researchers note that fisheries that yield more than 80 % of the global catch have not been formally evaluated. For their study, they used species, life history, annual catch, and fishery development data from thousands of fisheries globally to measure the status of thousands of fisheries in the world's oceans. They found that small non-assessed fisheries are in worse condition than their assessed counterparts. However, they estimate that large unevaluated fisheries may be performing nearly as well as their evaluated counterpart. Although the researchers calculate that both small and large stocks continue to decline, 64 % of the non-assessed stocks could yield increased sustainable catches if they are rebuilt following, to some degree, published suggestions [26, 27] that indicate that recovery of these stocks would create and increase in abundance by 56 % and fisheries yields by 8–46 %.

3.3.4 The Impacts of Global Warming/Climate Change on the World's Food Supply

Food crops, livestock, and fisheries are highly climate dependent to support conditions that optimize growth and maintain their health. On the basis of Intergovernmental Panel on Climate Change (IPCC) reports, international, governmental, and NGO environmental protection groups worldwide continue to assess the effects of climate change on the global food supplies. They are proposing policies that would assure food availability, directly or indirectly, for existing and future larger populations in nations with food deficits. The changes in climate that can cause a decrease in global food supplies are partly the results of natural variations in climate as well as climate changes abetted by human activities that power global warming.

Global warming is likely the core cause of extreme weather conditions. The harmful effects of climate change will be magnified in less developed and developing countries that have high population growth because many do not have the resources (natural and/or economic) to adapt to changes by applying management systems for crops (e.g., improved irrigation efficiency), livestock (e.g., disease control), and fishing (e.g., quota control on declining fish populations).

3.3.4.1 Crops

Crops grow with best yields and nutritious values when there is a balance between nutrients, water availability (moisture, precipitation), exposure to sunlight, temperature (daytime and nighttime), and CO_2 levels. Crop yield and quality will be degraded by the frequency, intensity, and duration of extreme weather conditions (heat waves, drought, tropical storms, and floods). They will also suffer as weeds, pests, and diseases migrate into formerly cooler agricultural ecosystems as warmer temperatures evolve. Adaptation to changing conditions that affect crop development is the key to maintaining planting and harvesting as the effects of global warming/climate are felt, over time, with more intensity, in ever-expanding regions.

The problem of seasonal drought or extended drought may be mitigated in two ways as noted in Chap. 2: (1) capture water during times of precipitation and storing it in reservoirs or in prepared underground caverns and (2) adopt water management measures to lessen impacts of water shortages (e.g., drip irrigation).

As some agricultural ecosystems become warmer over time, farmers can use seeds from productive areas that are normally at these warm conditions to grow crops in the new setting that are tolerant of higher daytime and nighttime temperatures. Also, farmers can diversify crops to accommodate changing temperature and precipitation or water availability conditions. They can adapt methods that have been used to control invasive weeds, pests, and disease in other areas where the methods have been used and have been effective. Certainly, traditional hybridization, MAS hybridization, and genetic engineering can focus on breeding seeds for specific crops that are resistant to pests and disease, and tolerant of heat, frost, drought, and submergence. This takes time to develop and test so that research now into the needs of evolving and future crop agriculture is a contemporary continuing process. As discussed earlier, nutrient replacement and weed control are managed by the use of agricultural chemicals and, where used, these should be applied only in amounts sufficient to support optimum crop yield and nutrition value. This will limit runoff of these chemicals and the damage they can do to ecosystems.

3.3.4.2 Livestock

Extreme weather events that are strengthened by global warming/climate change can be damaging to livestock in many ways, some that can be mitigated and some that cannot. Heat stress can reduce animal fertility, reduce milk production, and over time can increase vulnerability to disease. Heat waves lead to heat stress especially where there are high-density commercial operations. Whether dealing with cattle, chickens, pigs, or other meat providers, increased ventilation in barns (e. g., with fans), or at enclosures that are covered can alleviate the heat stress on animals and lessen their susceptibility to sickness and disease. Clearly, if there is a less densely packed population of food animals in an enclosure, there is a lesser chance of disease spreading among them. Shade trees in open areas where animals graze can reduce heat stress and serve to maintain healthy and disease-free food animals. Where extended drought becomes a problem, it threatens pasture and feed sources (e.g., corn) for grazers (e.g., cattle, sheep, goats) so that stored feed (and water) supplies have to be available to sustain them until the drought breaks. Lacking this, food animals die as has happened to millions of livestock in recent years, for example, in the USA, Argentina, and Australia. As warming trends advance into livestock-producing areas, it will be necessary to raise animals that can thrive under changing environmental conditions and to continue breeding programs that improve animal abilities to respond to environmental changes [28].

Global warming/climate change will surely bring about an increase in parasites and diseases (pathogens). This can be the result of warmer winters and early springs with increased rainfall and moisture that allows carriers of disease to flourish and invade livestock habitats. As just suggested, a reduction in density of penned animals should help limit the spread of an illness among them. When a veterinarian diagnoses sickness in an animal, it can be treated by a prudent use of antibiotics. Environmentally, this is preferred to loading the daily ration of water and feed for all animals in a herd or feedlot with antibiotics. The discreet use of these drugs limits their entering ecosystems via runoff or field disposal of wastes that can harm life forms in water and on land. This lessens the chances for a mutation of pathogens to antibiotic-resistant strains.

3.3.4.3 Fisheries

Global warming/climate change is causing a rise in fisheries water temperatures in oceans, estuaries, rivers, and lakes. This is affecting major fisheries now and will affect them more in the future [29]. Aquaculture in ocean enclosures or in inland waters is also exposed to the effects of warming waters. This, coupled with overfishing, will affect the global food fish supply. Fish and shellfish are very sensitive to temperature and seasonal climate changes. They have a survival temperature range. Out of this range, reproduction is reduced or will not take place and survival of young fish that hatch is endangered. As a result, where water temperature is warming out of the survival range, fish and shellfish migrate to cooler water. For example, cod in the north Atlantic need a temperature of <54 °F (~12 °C); sea bottom temperature >47 °F (~8 °C) harms reproduction and endangers the survival of young cod [30]. Hence, the cod migrate northward. Similarly, many other marine species have shifted northward as air and ocean water temperatures rise. For example, from 1982 to 2006, pollock migrated 30 miles northward and the snow crab 60 miles northward [30]. When fish/shellfish migrate, they can face a hostile ecosystem because there is competition from indigent fish and shellfish for food and other natural resources that sustain them.

The effects of warmer ocean waters on global fisheries also increase the possibility of disease that can endanger a myriad of aquatic life forms in marine ecosystems. For example, in the Pacific Northwest of the USA, warmer waters may affect salmon biorhythms and increase the potential for disease. Together these can cause a decline in salmon populations that in turn will hurt local economies. A lobster shell disease is sensitive to a warmer temperature change and already affects southern New England catches [30].

The increased partial pressure of CO_2 in the atmosphere (280 ppm preindustrial revolutions to almost 400 ppm in 2013 (a greater than 41 % increase) reached a critical concentration, and the ocean chemical environment is changing toward a less basic water toward an acidification condition. Acidity is measured on a pH scale of 0–14 with a value of 7 being neutral. Values less than 7–0 represent increasing acidity. Values from 7 to 14 represent increasing causticity (basicity). The ocean has a pH of about 8.1 but has been changing toward less than 8.1. The less basic waters in the oceans inhibit the building of shells (by precipitation). This puts the continued existence of many shelled life forms at risk. As the trend toward acidity continues into the future, not only will many shelled marine animals suffer but coral reefs will suffer as well. Reduction of coral reef ecosystems because of a decline in coral reef building caused by ocean acidification will result in a loss of habitats for spawning and growth of fish. This retards the rebuilding of fish stocks that we know have been greatly reduced by overfishing. Add to this, the effect of warming waters and fish/shellfish migration plus the increasing trend of ocean water toward more acidification and the global outlook for food fish is a decline of fish populations. There is no easy solution to these problems other than stemming the emissions of CO_2 and other greenhouse gases into the atmosphere.

This is an international problem that will likely be magnified in regions with high population growth and in developing countries that do not have enough economic resources to cope with it. It is a special problem when they do have available resources but do not use them to stem the emissions of CO_2, as is the case for China, and therefore adapt to the needs of the greater global society or even their own citizens. Here, the developed countries have to step in as necessary when requested to do so and help the developing world with economic assistance and technical know-how.

3.4 Afterword

We have discussed the problems of water and food needs of the earth's populations now and in the future when the populations will increase by more than a third from 7.2 billion people in 2014 to 9.7 billion people in 2050. We have reviewed the corrective strategies that can be adopted and acted on by governments with the will to solve the problems of water accessibility and food availability using their own resources and supported by added financial and technological resources from international institutions and NGOs. To shelter the growing global populations at locations where they can be secure from natural disasters in their homes and at work is to satisfy a third basic human right. This is the topic of the following chapter.

References

1. Food and Agricultural Organization of the United Nations (2011) The state of the world's land and water resources for food and agriculture—SOLAW—Managing systems at risk. Rome, and Earthscan London 294 p
2. Food and Agricultural Organization of the United Nations (2012) FAO statistical yearbook 2012. World food and agriculture. Rome 362 p
3. Grimaraes E, Ruane J, Scherf D, Sonnino A, Dargie J (eds) 2007. Marker-assisted selection. FAO of the UN. Rome, 441 p http://www.fao.org/docrep/010/a1120e00.htm Accessed 14 Mar 2014
4. Food and Agricultural Organization of the United Nations (2010a) Current status and options for crop bio-technologies in developing countries. Int'l Technical Conference, Guadalahara, Mexico 67 p
5. Gupta PJ (2009) Bt and cotton marker-assisted selection for crop improvement in India. Message 2 of the FAO e-mail conference on learning from the past www.fao.org/biotech/logsc16/170609.htm Accessed 14 March 2014
6. Rigor A (2009). Experience from Philippines rice. Message 42 of the FAO e-mail conference on learning from the past www.fao.org/biotech/logs
7. Xu Y, Crouch JH (2010) Marker-assisted selection in plant breeding: from publication to practice. Crop Sci 48:391–407
8. Gamuyao RM Chin JH. PariascaTanaka,J, Pesaresi P Catausan S Dalid C Slamet-Loedin I. Tecson-Mendoza EM Wissuwa M Heuer S (2012) The protein kinase PSTOL1 from traditional rice confers tolerance of phosphorus deficiency. Nature vol 488:535-539
9. Siegel FR (2001) Environmental geochemistry of potentially toxic metals. Springer, Berlin, p 218
10. Ronald PC (1997) Making rice disease resistant. Scientific American, USA, pp 100–105
11. Potrykus I (2001) Golden rice and beyond. Plant Physiol 125:1157–1161
12. Karyeija RF, Gibson RW, Valconen JPT (1998) Resistance to sweet potato virus disease (SPVD) in wild East Africa Ipomoea. Ann Appl Biol 132:39–44
13. Wikipedia (2012) Genetically modified crops. See table that lists genetically modified crops
14. U.S. Department of Agriculture, Economic Research Service (2013) Adoptation of genetically engineered crops in the U.S. (1996–2013), Washington, D.C. www.ers.usda.gov/data-products/adoption-of-genetically-engineered-crops-in-the-us/recent-trends-in ge-adoptation. aspx. Accessed 14 March 2014
15. Singh OV, Ghai S, Paul D, Jain RK (2006) Genetically modified crops: success, safety, assessment, and public concern. Appl Microbiol Biotechnol 71(5):598–607
16. SANET(2006) GM ban overdue, http://lists.ifas.ufl.edu/cgi-bon/wa.exe?A2=ind0602&L= sanet-mg&P=3448
17. Mellon M, Rissler J (2003) Environmental effects of genetically modified food crops—recent experiences. Union of Concerned Scientists, Cambridge, p 4
18. Despommier D (2010) The vertical farm: feeding the world. St. Martin's Press, New York, p 302
19. Derpsch R (2012) A short history of no-till. No-till, a series of papers. http://www.rolf-derpsch.com/notill.htm

20. Tripplett GB, Dick WA (2007) No tillage crop production: a revolution in agriculture. Agronomy J 100:S-153–S-165
21. Qadir M, Ghafoor A, Murtaza G (2000) Amelioration strategies for saline soils: a review. Land Degrad Dev 11:501–521
22. Looft T, Johnson TA, Allen HK, Bayles DO, Air DP, Stedtfeld RD, Sui WJ, Atedtfeld TM, Chai B, Cole JR, Hashsham SA, Tiedje JM, Stanton TB (2012) In-feed antibiotic effects on the swine intestinal biome. Proc Natl Acad Sci, doi: 10.1073/pnas.11202381109, Washington, D.C., 6p
23. Chee-Sanford JL, Mackie RI, Koike S, Krapac G, Yu-Feng L, Yannarell AC, Maxwell S, Aminov RI (2001) Fate and transport of antibiotic residues and antibiotic resistant genes following land application of manure waste. J Environ Q 38(03):1086–1108
24. Nkrumah JD, Okine EK, Mathison GW, Schmid K, Li C, Basarab JA, Price MA, Wang Z, Moore SS (2006) Relationship of feedlot feed efficiency, performance, and feeding behavior with metabolic rate, methane production, and energy partitioning in beef cattle. J Anim Sci 84:145–153
25. Food and Agricultural Organization of the United Nations (2010b) The state of world fisheries and agriculture, Rome, 197 p
26. Worm B, Hilborn R, Bau JK, Branch TA, Collie JS, Costello C, Fogarty MJ, Fulton EA, Hutchings JA, Jennings S, Jenssen OP, Lotze HK, Mace PM, McClanahan TR, Minto C, Palumbi SR, Parma AM, Ricard D, Rosenberg AA, Watson R, Zeller D (2009) Rebuilding global fisheries. Science 325:578–585
27. Costello C, Ovando D, Hilborn R, Gaines S, Deschesnes O, Lester S (2012) Status and solutions for the world's unassessed fisheries. Science 338(6106):517–520. Online 27 Sept. 2012. doi: 10.1120/science.1223389
28. Thornton PK (2010) Livestock production: recent trends, future prospects. Philos. Trans. The Royal Soc. B27 365(1554):2853–2867
29. Daw T Adger WN Brown K Badjeck M.-C (2009) Climate change and capture fisheries: potential impacts, adaptation and mitigation. In Cochrane K De Young C Soto D Bahri T (eds). Climate change implications for fisheries and aquaculture: overview of current scientific knowledge. Fisheries and aquaculture technical paper 530, FAO, Rome, pp 107–150
30. Karl TR, Melillo JM, Peterson TC (eds.) (2009) Global Climate change impacts in the United States. US Global Change Research Program. Cambridge University Press, Cambridge 188 p

Chapter 4
Shelter: Proactive Planning to Protect Citizens from Natural Hazards

4.1 Introduction

There is space on the Earth to locate 3 billion more people than inhabit the Earth in 2013. But locating space that is safe from natural and anthropogenic hazards and the events they trigger or that can be made safe by investment, sound engineering, retrofitting, human innovation, and common sense can be problematic. There are surely limits to how much more population megacities can accept without becoming unmanageable. The reasons are that in addition to establishing safe sites for shelter (plus workplaces and infrastructure), there must be accessibility to foodstuffs and clean water, and opportunities for employment to support a tax base that can provide social services (e.g., healthcare and education) for a community. This chapter will focus on the physical aspect of site selection and technology that can protect growing, high density populations from natural hazards.

4.2 Foresight in Site Selection from Studies of Past and More Recent Hazard Events

Each time there have been earthquakes, volcanic eruptions, massive flooding, damaging mass movements, violent storms, and other natural hazards historically and in modern times, scientists (e.g., geologists, engineers, chemists, and biologists) glean new information from observations and measurements of the events. These data include, for example, the frequency of an event at a location, the magnitude and type of force that is released, the flow paths that fluids follow, and the response of structures to the energy emitted. Primary hazards can trigger secondary events (e.g., tsunamis and floods) that can be more damaging to people, structures, and local and regional ecosystems. Evaluation of such information can guide scientists and building code officials in the future to implement adaptations that make existing

© The Author(s) 2015
F.R. Siegel, *Countering 21st Century Social-Environmental Threats to Growing Global Populations*, SpringerBriefs in Environmental Science,
DOI 10.1007/978-3-319-09686-5_4

sites less dangerous in the future for people and their environments. Conversely, it can put planned residential, commercial, agricultural, and industrial development and supporting infrastructure on hold unless developers agree to incorporate the latest available technologies and materials that will provide planned structures with resistance to the destructive energy of natural hazards.

Subsequent sections in this chapter consider first how citizens can be discouraged from inhabiting zones at high risk from one or more natural hazards and secondary events they cause. A second topic discussed is what municipalities and the application of technology can do protect growing population centers and edge cities on six continents from occasional destructive hazards that have impacted them in the historical and recent past. The third theme is an assessment of the preparation necessary to cope with the immediate effects of a major disaster on citizens and its traumatic effects on a stunned and suffering population [1].

4.2.1 Earthquakes

Earthquakes occur regularly and there are thousands recorded each year. The movements generated by earthquakes are shaking, jarring, or rolling motions or combinations of these with varying force. Most earthquakes have magnitudes that are imperceptible to moderately perceptible (1–3 on the Richter scale) but not damaging. Others are alarmingly perceptible but generally cause only minor damage to property but are not often a threat to life (4–5 on the Richter scale). The Earth suffers a few earthquakes each year of great magnitude, some on land and some in the ocean, during and after which people are killed and property is destroyed (6 and higher on the Richter scale). Earthquakes do not kill the people. Collapsing buildings or infrastructure kills people. Secondary events, such as tsunami generated by an earthquake, kill people and destroy property (Sumatra, 2011) or those that result from violent earthquake movement such as fires from overturned stoves (Tokyo, 1925) or ignited broken gas lines (San Francisco, 1906).

Then we have a tsunami such as the one that followed an earthquake of great magnitude (9.0) off the northeast coast of Japan in 2011 that not only destroyed entire coastal towns and villages, killed 15,883 people with 6,150 injured and 2,643 missing according to the Japanese National Police Agency, but also wreaked havoc at three nuclear reactors of the Fukushima Daiichi power facility. The tsunami reached 15–23 m high in some coastal stretches as it moved shoreward, easily overrode a "protective" 5.72-m-high seawall, and rolled up on shore wrecking the nuclear power facility and shutting down the primary system that injected waters to cool the spent fuel pools where used uranium pellet rods were stored. The tsunami destroyed the seawater intake system and seawater accessed the emergency diesel generators and switchback rooms, cutting the power to the water pumping units that were designed to continue to circulate water that cooled the reactor radioactive materials if the primary systems failed. This resulted in a complete loss of cooling capacity and meltdowns and explosions that released deadly radiation into the atmosphere and rendered an important swath of land in northeast Japan unusable. As a result of reactor meltdown that breached the protective foundation, there was a release of radioactive waters. Some of these waters were still being captured during the fourth quarter of 2013 and stored in tanks until a solution is found that can reduce the radiation to an acceptable level for release. However, 400 tons of radioactive waters are discharging into offshore coastal ecosystems daily with ideas but no actions to stem and stop this flow.

The land-use planning error here was in site selection from not evaluating earthquake and tsunami hazard potential because engineers did not do a thorough literature search that would have alerted them to possible problems. There were engineering errors in plant design by not having a backup seawater intake option, and by not housing generator/pumping capability in waterproof facilities. In 2001, scientists indicated that an earthquake with a magnitude estimated at 8.3 in 869 AD off of Sendai Bay, generated a tsunami calculated to have been about 8 m high [2]. They estimated a frequency of occurrence at 1,000 years. These data alone should have alerted the managers of the nuclear power facility that their 5.72-m-high seawall was very susceptible to be topped in this earthquake/tsunami prone zone. Thus, if Tokyo Electric Power Company engineers and consultants were up to date with a pertinent scientific journal dealing with disasters, they had an alert 10 years earlier that the seawall height should be increased. Coastal engineers erred by not setting a more realistic "safe" height for the protective seawall for the Fukushima Daiichi nuclear power station and coastal population centers based on published historical data [2]. Here, then we have not only a triggered tsunami but also a tertiary happening, likely abetted somewhat by human failings that killed people, destroyed towns and villages, and poisoned the environment with radiation for some time in the future. An extensive review of the Fukushima Daiichi accident (more than one year later) was published in June 2012 number of the journal Elements. The magnitude 9.3 earthquake that generated a tsunami off the northeast coast of Sumatra, Indonesia in December 2010 and that also overran coastal areas in Thailand, Sri Lanka and India and other nations killed more than 240,000 people had neither seawall protection nor a tsunami alert and warning system.

4.2.2 Volcanic Eruptions

There are more than 550 active volcanoes on Earth. This number is based on historical records for the past 10,000 years. In any decade, 160 volcanoes are active to varying degrees. When one erupts, it may be effusive and emit lava that flows downslope following the topographic lows. An eruption may be explosive and blast glowing tephra, cinder, and ash, up into the atmosphere together with the pulverized rock materials it has breached or the eruption may explode as lateral blasts that ravage the areas reached by the radiative heat, gases, and solid materials. These primary events can kill people especially when what is emitted is a nuee ardente (a glowing cloud) of super-heated ash and gas that flows downslope burying people and their cities and/or sucking oxygen out of the atmosphere to support the burning, thus asphyxiating and encasing in ash all life forms in its path. Classic examples are from Mount Vesuvius in 79 AD that destroyed and buried the towns of Pompeii and Herculaneum and killed 3,000 people, and from Mount Pelee on the Caribbean island of Martinique that rolled over the capital St. Pierre in 1902 killing 30,000 people. Great volumes of ash ejected high into the atmosphere was a primary volcanic hazard that disrupted air traffic globally in 2010 when the Eyjafjallajokull volcano in Iceland erupted. The masses of ash drifted south and east across Great Britain and Ireland to Denmark, Norway, Sweden, Finland, and France closing airports and disrupting standard flight paths to and from these nations and other nations for several days. Great emission of ash from Mount Pinatubo in the Philippines during 1991 caused a lowering in global temperature by about 1 °C for a year as the ash moved by directional winds circled the earth and prevented a portion of sunlight from reaching its surface. Depending on the topography, people can generally get away from a lava flow such as at Mt. Kilauea, Hawaii, although property and infrastructure in the path of a flow is destroyed.

Many deaths and property losses that result from volcanic activity are from secondary events such as hot mudflows called lahars that can roll over and engulf cities, burying them and their inhabitants under many feet of mud. A major event of this kind happened when the volcano Nevado del Ruiz in the Columbian Andes erupted in 1985 and melted its ice/snow cap. The melt mixed with the materials on the slopes of the volcano and set loose a mudflow that moved downslope at high speed following the topographic lows into the valley and onto and over the town of Armero killing an estimated 25,000 people. In some cases, masses of mudflow into streams and rivers, creating dams thus flooding and disrupting ecosystems upstream. Fires ignited by glowing tephra and hot ash falling to earth ravages structures and disrupt ecosystems by igniting forest fires and poisoning waterways. Thus, before situating people in an active volcano area, all potential volcanic hazards, primary or secondary, have to be evaluated for risks to them and their possessions. As with earthquakes, scientists assess historical and contemporary records to estimate the type of eruption, the frequency of events, the force they may release, the paths volcanic matter is likely to follow, and the triggered events and their geographic reach.

4.2.3 Floods

In the past few years, severe flooding affecting large populations seems to have been more frequent, more severe, and longer lasting in countries on six continents. This is likely supported to a marked degree by global warming because of warmer ocean waters with more surface area from sea level rise that evaporate more moisture into the atmosphere. This loads clouds with moisture that yield more rain lasting for longer periods of time. The runoff fills streams and rivers in volumes that cannot be contained in their channels. The excess waters overflow onto natural flood plains and then beyond to the 100 year flood plain and farther yet in some areas where the flood levels put a 500 year flood plain under water. If people in cities or suburban and rural areas are unaware of developing flood conditions, they may be caught in the flood waters and suffer injury or drown and structures and infrastructure damaged or destroyed. However, if people are warned in time, they may be displaced and their homes and possessions lost, but they will be alive and that is the important thing. Agricultural fields may be submerged and crops lost. Rapidly, flowing flood waters can erode and undercut valley walls triggering landslides thus putting people, structures, and infrastructure set above the valley walls at risk of slipping downslope and thus putting people, structures, and infrastructure in the land beneath at risk. The rainwater itself can load slopes with weight that increases stress that slope materials cannot sustain and also provides lubricant in slope soil/rock masses thereby favoring landslide movement. Masses of landslide materials falling into rivers can act like dams further exacerbating flood conditions upstream and subsequently releasing ravaging waters downstream when the pressure exerted by the restrained waters breaches the temporary dams. Another secondary event that can develop after great floods is disease (e.g., typhoid and cholera) if sanitation systems are damaged or destroyed.

4.2.4 Mass Movements

Landslides, rock falls, subsidence, and collapse comprise the environmental geologist's mass movement category. They occur when the strength of a rock resisting stress is overcome by the pull of gravity. In the case of landslides, slopes greater than 15° are susceptible to failure mainly when rain or snow melt seeps into soils and sedimentary rocks. This increases the weight and gravitational stress on the slope and decreases the friction within slope materials further reducing slope resistance to failure. Rock falls from high angle cliffs are common in temperate climates. They are often the result of water that seeps into cracks and crevasses in rocks that freeze during the night with the expanding ice wedging the cracks farther part. As wedging continues over the course of years, the pull of gravity overcomes the strength of the weakened rock and masses of rock detach and can cause additional rocks to detach as they cascade down smashing against the rock face. In both cases, the likelihood of deaths is low but property damage to structures, roads, and utilities can be extensive.

Subsidence of an extensive area of land or collapse of surface materials that cause sinkholes can also be destructive to structures, roads, and utility systems but is rarely an event that causes injury or death. Again, the pull of gravity is the driving force but the process that brings on the mass movement is unique. Fluids in sedimentary rocks in the subsurface, perhaps aquifers, perhaps sedimentary rocks charged with oil and natural gas, have added strength from what is called buoyancy pressure. When the fluids are withdrawn and are not recharged, this buoyancy pressure is lost so that the weight of the overlying rock masses can compact the underlying rocks. Depending on the strength of the overlying rocks, there may be no subsidence or there can be a lowering of the earth's surface locally or over an extended area. A classic case of subsidence is Mexico City where increasing withdrawal of water from underlying aquifers over time to satisfy a growing city population caused a lowering in some areas of up to 8 m. Weight of structures adds to the compaction and subsidence. The Palacio de Bellas Artes is a massive building of some 60,000 tons that was completed in 1906. It has subsided more than 3 m so that theater-goers today walk down marble stairs to reach the entrance that once was at the ground floor. Extraction of oil near Long Beach Harbor, southern California began in 1938 and by 1958 subsidence reached 9.5 m and extended to Los Angeles Harbor. Oil wells and pipelines were damaged as were pier facilities, and harbor and city infrastructure. More than $100 million was necessary to repair the damage. As important were the calculations by engineers that further extraction of petroleum (and the 8 barrels of briny water that were extracted with each barrel of oil) would lead to subsidence that could reach 22 m. To stabilize the situation, the extracted waters (brines) plus added water from shallow aquifers were injected into the oil bearing strata to re-pressurize them. There has been rebound of 30 cm in some areas of the oil field. Similarly, near-surface underground coal mining worldwide has caused subsidence as well with damage to property and infrastructure because not enough support pillars were left in place by miners, or pillars themselves were removed for their coal content as exploitation of a mine was ending.

Sinkholes occur as a result of collapse. This happens most often where the subsurface rock is limestone and where there is a high water table. Water slowly dissolves the limestone as it flows through the rock. Over the course of millions of years a void or underground cavern forms and continues to grow. When the pull of gravity on the roof rock that was undermined by solution exceeds the strength of the unsupported roof rock, it collapses and a sinkhole is born. Structures or infrastructure overlying a collapsing sinkhole will be destroyed but there is only a small likelihood that people will be injured or die. Collapse has also occurred where there has been near-surface mining, especially for coal when support columns were themselves mined as noted in the previous paragraph.

4.2.5 Extreme Weather Events

Extreme weather events seemingly increased in number and violence during the past several years. They have killed and injured people and have damaged or destroyed property. Some of this is the result of secondary events. High-energy tropical storms (hurricanes, typhoons, and monsoons) are events during which high velocity directional winds can tear apart buildings and propel fragments like missiles that can kill and injure citizens. The storms can generate surges that carry seawater onto land with forces that destroy coastal structures as well as invade coastal agricultural fields and pollute them with salt water. The reach inland of storm surges during the past few decades seems to be greater than in the past.

Heavy and sustained rains during the extreme storms can cause flooding and load slopes with water thus priming landslides and mudflows that endanger people, property, utilities, and other infrastructure installations. According to the United States meteorological reports, violent tornado activity in the central and southern USA has increased in recent years.

Extended heat waves have been killers vis a vis 34,000 who died (14,000 in France) during a heat wave in Europe, August 2003. Long lasting droughts also kill livestock and damage or ruin crops as happened in Midwest USA during August 2012. This is an economic negative as the price of farm products increases, including animal feed, and this causes a price increase in foodstuff for the consumer. Similar drought and heat related events have occurred in Argentina and Australia in the past few years affecting livestock and crops important to the countries' export economy.

Extreme weather events such as noted above are recurring events the effects of which have surely intensified because of global warming/climate change.

Evaluation of natural disasters discussed in this section, both historical and contemporary, allows governments and scientists to develop concepts that can protect people and property. They bring these concepts into practice before a settlement is built in a hazard zone for growing populations, or before a disaster slams into existing population centers. The following two sections will deal with this approach.

4.3 Sociopolitical and Economic Methods to Discourage Habitation of High Hazard Zones

The social approach is to educate citizens of the potential problems for themselves, structures, and infrastructure that can result from human occupation of yet unoccupied high hazard zones defined by scientific analysis. This is done by offering discussions between would be inhabitants and risk assessors and scientists who have good communication skills and present easily understood imagery that shows through actual case studies what have been the results in like areas when natural hazards have impacted a population center. They can give the probability of such an event taking place during given periods of time, emphasizing for some events the potential future effects of global warming/climate change on their estimates. This is especially important when considering occupation of flood plains by determined individuals. What have been defined as 100-year flood plains in the past, that is, the level of flood waters that is predicted to overflow stream/river banks once in 100 years, is likely to be superceded in a lesser period of time and encroach on higher elevations in areas where climate change brings more precipitation. The same approach can be used when considering requests for human settlement in other high hazard event zones. If this does not deter citizens, the government can step in.

The political solution to protect people from themselves is to pass zoning legislation that prevents human habitation of zones identified as dangerous to people and property. The decision to do this is based on historical and modern observations and measurements that show that there is a high probability of a hazard event occurring there during citizens' lifetimes. If a government does not want to employ zoning, it can refuse to provide infrastructure to the high-risk zones and use high taxes as a deterrent to occupancy. Governments may not want to impose either of these methods to discourage human settlement of these known high hazard zones so that additional economic means may be used.

Economic constraints can prevent living in unoccupied zones that are at high risk of being hit by a natural hazard within citizens' lifetimes. Banks can refuse to offer mortgages to build in such areas or can require a high initial payment and secure collateral for 100 % of the remaining loan because the land and structure are not security worthy. If citizens receive loans under these conditions, private insurers or government insurers can refuse to write policies that cover lives and property in the high-risk areas.

If people persist and have the economic means to build and live in areas at high risk of suffering from natural hazards, then they can hope that there will be means available to protect them from the threat and impact of a hazard event. Communities established before scientists defined their locations as high-risk hazard zones have options as to how they will react in the future if the government will or will not implement programs that can assure their safety. These are discussed in the following section.

4.4 Practices to Protect Citizens/Property from Occasional Battering by Natural Hazards

The death and injury and damage and destruction that can result from contact with high-energy natural hazards can be damped to a lesser or greater degree following recommendations in geologists' reports: (1) application of engineering and technological advances based on evaluation of former like events; and (2) disaster response planning. This is true for established, densely populated communities, and for the construction of new communities that are built to accommodate demographic changes in many countries.

4.4.1 Building Codes

Building codes define several parameters for construction that are designed to make structures safe for use by citizens. The codes change as more is learned by scientists, engineers, and risk assessors from evaluation of damage and destruction data after an event and/or triggered happening of a given magnitude has impacted a community. In the case of areas under the threat of earthquake activity, the codes define the earth materials that are best to build on and how a building's foundation must be tied to the underlying materials. They are explicit in the materials that can or cannot be used in construction and how these must be tied (bolted, welded) together to give the maximum structural stability to a building during an earthquake. Building codes also give the best ways to retrofit structures to make them more resistant to earthquake motions. In the case of high rise structures, the codes may recommend or require special foundations that may be on springs or paddings that allow a building to absorb some of the earthquake motion energy and thus increase resistance to damage or collapse. Similarly, they may require moveable weights on rooftops that are controlled by computerized systems that receive seismic signals giving motions generated by an earthquake and instantly shift the weights so as to counter motions from seismic waves.

In volcanic areas where explosive eruptions send incandescent tephra and ash into the atmosphere but where construction may be allowed, building codes may require that the construction materials be nonflammable so that hot ash falling on buildings will not ignite them. They may also require that structures have high pitched roofs so that extended ash falls will not build up to unsustainable weights, especially when they become wet and could cause collapse.

When building codes are brought up to date regularly on the basis of experience with any type of natural hazard, people and property will be less subject to injury or death, and damage or destruction of their property. However, no building code can protect structures or infrastructure in the path of lava flows nor protect citizens and their property in the path of major floods. Protective structures as described in the following section can be effective in some cases during lava flows and in most cases during floods.

4.4.2 Protective Infrastructure

Protective structures are most often associated with water in the environment. Dams, levees, and dikes offer protection against flooding in populated areas that would otherwise be inundated and subject to life-threatening conditions and damage. This depends on the amount of water entering and overflowing a channel, the velocity with which it moves through an area, and how quickly flood waters ebb. In response to the potential of flooding, a municipality may choose to deepen, widen, and/or straighten a stream/river channel where the waters pass through populated communities. This allows a channel to carry more water and move it more quickly through an occupied or agriculturally productive land area at risk of submergence.

Similarly, seawalls are built to protect coastal zones from tropical storm wind-driven storm surges that might cause injury or death to coastal populations, damage buildings, and infrastructure, or simply inundate cities or villages or agricultural fields in their path. Shanghai, China has 500 km of built-up embankments and a protective 500 km seawall with a height that has been increased according to happenings in the past and is now more than 6.6 m high. The Netherlands has invested billions of dollars to protect its inland areas and ports from inundation by North Atlantic storms. The Dutch build and maintain dikes and in some cases have installed hydraulic/electric-driven walls that can be raised into place to protect its ports. Buried metal barriers in the Thames River channel can be raised hydraulically to protect London from flooding by tidal bores-driven upstream during major storms. As noted earlier, seawalls have been built to protect populations and infrastructure from tsunami events as well as from storm surges.

In many parts of the world, landslide damage is a great economic burden. In order to minimize landslide activity, engineers develop options to prevent water with its destabilizing weight and lubricating capability from building up in susceptible slopes. In some cases, depending on geology and engineering reports, partially perforated pipes are strategically placed in slopes to capture in seeping rain or snow melt and direct it to safe capture areas away from a slope. This minimizes the possibility of landslide activity. In other areas, slopes with a high potential for damage to critical infrastructure (e.g., to railroad tracks and electrical transmission lines) if they fail are sealed with concrete-like material to prevent water from seeping into them. People and infrastructure in densely populated centers are protected from rockfalls by chain link barriers bolted into the face of the potentially dangerous rock mass.

In areas of Japan with a history of eruption and lava flows, attempts at protection is provided by the installation of barriers grouped at angles designed to deflect and redirect lava flows so that they move away from inhabited locations.

Proactive protective planning and implementation has saved populations from injury and death and has lessened the probability of damage to residential, commercial, industrial, and agricultural activities that are important to society. However, the chances to maintain conditions that enhance the safety of people can be improved by predictions of where, when, and with what magnitude hazard events will occur.

4.5 Prediction

Unlike many industrialized nations that can invest in the construction of protective infrastructures such as those noted in the previous section, many countries do not have the economic capability to invest in them. Thus, the ability to warn people of an imminent natural disaster that allows them to evacuate areas with critical documents and whatever else is dear to them is of utmost importance.

4.5.1 Earthquakes

Earthquake prediction is in active research but is not yet a working tool. Although there have been a few successes in short-term predictions, they have been rare and sometimes followed by devastating failures. For example, during the winter of 1975, China ordered the evacuation of Haicheng, a city of about one million people after geological scientists and other sources over a period of a few months reported changes of elevation in the nearby Bohai Sea and in groundwater levels plus widespread anomalous animal activity and regional increases in seismic foreshock activity. The day after the evacuation, the city was lay waste by a major earthquake but only about 2,000 died with about 27,000 injured. The Chinese estimated without the prediction and evacuation, more than 150,000 people would have been killed or injured. However, in July 1976, an earthquake with a magnitude of 7.6 hit Tangshan, China, also a city of about one million people, without warning because there were no precursors and no evacuation. The death toll was 240,000 people with 164,000 injured. At best, geological scientists specialized in earthquake seismology and using additional historical and observational data and physical measurements can make general earthquake predictions but not within a time frame that would allow people to evacuate high-risk areas. For example, the United States Geological Survey has predicted that one or more earthquakes of magnitude 6.7 or higher will happen in California within 30 years with this prediction having an estimated 99 % probability of accuracy. This is an interesting topic for research scientists but with no real substance for the California public. The science of earthquake prediction continues to be a necessary research topic that will ultimately save lives. This will be the case when scientists are able of give warning in days or weeks ahead of an event that would allow people time to prepare or to evacuate.

4.5.2 Volcanic Activity

Volcanoes do have observable and measurable precursors that allow a reasonable degree of predictability to allow evacuation of citizens from high-risk hazard zones. For example, wisps of smoke from a volcano indicate activity. Gaseous emissions from a volcano during the beginning stages of activity increase dramatically for sulfur dioxide as internal activity intensifies. Gases can be captured and measured by small planes passing over the volcano. A warming of soils or of lake waters near a volcano when compared with a baseline level determined when it was quiescent suggests that magma in the subsurface is moving toward the surface. This can be measured with thermometers or by time-line infrared aerial photographs. A change in ground deformation at a volcano as measured by tilt meters and telemetered to scientists can identify bulging where magma is rising, pushing surface material up and/or out. Measurements that show changes in magnetism and/or the pull of gravity can also be indicative of subsurface magma movement and an impending volcanic eruption. Anomalous animal activity has been observed prior to an effusive or explosive eruption. Finally, and most important, is the seismicity associated with a volcano. Monitoring the seismicity over time allows baseline levels to be established. What has been observed are changes in the frequency and intensity of seismic activity including harmonic tremors and/or increased long-period (low frequency) wave activity that are indicative of the probability of an eruption in 24 h. On the basis of monitoring a combination of such precursors, villages on or around volcanoes have been evacuated to the benefit of their populations.

4.5.3 Floods

Flood warning systems are in place in many areas where the hazard has been and continues to be a danger to life and property. They are in place even if physical controls of floods have been constructed or where changes of a stream/river channel have been engineered in order to move more water through a channel more quickly to minimize overflow or eliminate it. Flood warning systems consist of stream/river gauges in a drainage basin that measure the volume of water in various waterways in the basin and the velocity with which water is moving. Data from the gauges are telemetered to an installation where computers input them together with data on cross sections of channels where the measurements are recorded and on the declination and cross sections of flow channels downstream. With these data, meteorologists and hydrologists can accurately determine when flood waters will reach any area in the basin and the level flood waters will reach as they flow past a given point. Depending on the size of a drainage basin, citizens in areas that are susceptible to flooding may have days to prepare their fight against the water. This might include sandbagging, building up levees, moving furniture to upper floors in a home, encasing appliances or air conditioning/heating units with impermeable sleeves, or finally gathering important documents and treasured possessions and evacuating the flood zone until flood waters ebb.

4.5.4 Tsunamis

Tsunamis are sea waves that originate when there is a sudden displacement of a massive volume of seawater. This is most often the result of an earthquake when a large segment of the seafloor ruptures (faults) and is thrust up over the mass from which it ruptured or when the rupture causes a downward displacement of a seafloor segment. The vertical displacement of the seafloor can cause the generation of a tsunami. In the past, tsunamis have originated as well from great eruptive volcanic activity in the deep ocean, by giant landslides from the oceanic continental slope, and from meteorite impact, but these happenings are rare.

The displacement of the seawater can initiate a series of waves in all directions with their crests hundreds of miles apart (low frequency long wavelength movement) and initially with a low amplitude of 30–60 cm (~ 1–2 ft) that is hardly noticeable from the bridge of a ship plying the oceans. As these sea waves move across the ocean floor at velocities that average 500 km/hr (over 300 mph) and may reach 800 km/hr (500 mph). As they approach a coast, the waves builds up as the ocean bottom shallows and rise into a giant sea wave that can be 10 s of meters high. As an initial tsunami wave rides up onto a coastal zone, it can cause injury and death, and damage and destruction vis a vis northeast Japan, 2011 or Sumatra, Indonesia in 2010. This is exacerbated when the mass of water flows back into the ocean dragging bodies and rubble along with it. This is followed, perhaps hours later by sequential tsunami wave crests that roll into the coastal zone that can have a greater damaging force on structures weakened by the initial tsunami wave. Hundreds of millions of people worldwide live in coastal regions susceptible to tsunamis and this number will grow as global populations continue to expand to 2050 and beyond. Because of this, tsunamis are given extra attention in this chapter.

The problem of unannounced killer tsunami waves has a partial solution in the deployment of deep-ocean detection systems anchored to the ocean floor. These are specially equipped buoys that can rapidly identify that a major earthquake has occurred (seismic sensors), where it has occurred (GPS), changes in the height of seawater above the seafloor (sea bottom pressure sensors), and the rate at which waters are moving. These data are sent via telemetry to tsunami warning centers that process them with high-speed computers. This information and additional bathymetric (seafloor topography) data along the flow path of the sea waves are processed and warnings are sent immediately to coastal zones that predict how soon a tsunami will hit together with an estimate of the maximum height a tsunami wave will reach. It will also assess the status of secondary tsunami waves that follow and this entire process may take 5 min. Thus, for a tsunami produced not too distant in the deep ocean, the detection and warning system may not be effective.

About 80 % of tsunamis are in the Pacific Ocean and most of the tsunami detection buoys were initially in the Pacific monitoring activity mainly offshore USA, Japan, and South America. They were lacking in the southwest Pacific Ocean. This omission and lack of investment in a tsunami monitoring and warning system was surely responsible for many of the more than 240,000 killed by the 2011 magnitude 9.3 earthquake off the coast of northeast Sumatra, Indonesia and the subsequent sea wave that obliterated Sumatran coastal population centers and their inhabitants as well as populated coastal zones in 10 other nations. Tsunami detection buoys and a warning system are now in place in the rest of the Pacific region. When a massive 8.8 magnitude earthquake in the deep ocean off the Chilean coast occurred, it caused a 10 meter tsunami that hit Chilean Central Valley coastal area. Five hundred people died, half from the tsunami because the Chilean Navy did not immediately give a tsunami warning. However, hundreds of lives were saved because port captains in many coastal towns that were demolished gave the tsunami warning 23–30 min before the sea wave hit land so that people could evacuate inland to higher ground.

4.5.5 Extreme Weather Events

Meteorologists at storm tracking stations worldwide follow the beginning of the development of storm conditions over the ocean, determine a storm's energy, and track it as it moves over an ocean toward land. They monitor the velocity of its movement, the path it is following, and its energy as measurements change along its track. Storm trackers use data from satellites and specially equipped aircraft that can fly into the eye of a storm and measure wind speed, moisture, and areal reach. These data are input into computer programs to provide storm models and refine them as new data on changing conditions are added. With this information, meteorologists can warn populations days in advance and with reasonable accuracy where a tropical storm is likely to move onto land, when it will reach land and inland areas, the moisture (rain) it will deliver, and whether it is a low-level tropical storm or one that is a hurricane (typhoon, monsoon), its level of energy determined by its wind speeds, whether category 1, 2, 3, 4, or 5, the most damaging, and the potential effects of storm surges. This allows citizens to prepare their homes and businesses to get through the storm with minimum damage. It gives them time to gather important documents and precious possessions if it becomes necessary to evacuate the areas predicted to be impacted strongly by the storm. Much of these same data are used to predict where, when, and to what degree flooding will take place as a storm moves inland. In some cases, the rain data are used by geologists to alert areas to the potential for landslides. The onset of extreme heat conditions and of drought and how long one or the other will last is not readily predictable but warnings that may be issued at the onset of such extreme weather events can allow the public and businesses to prepare for them to the extent possible.

4.6 Pre-planning Post-hazard Responses: Citizen Care, Search, and Rescue Operations

No matter how careful siting is to avoid natural hazard zones, nor how good physical practices are to prepare population centers to withstand or minimize the impacts of natural hazards, the forces of nature can give rise to hazards that can overcome many if not most contra-hazard methodologies/technologies.

Thus, in an ideal situation, there will be careful planning and preparation to evacuate people when coupled monitoring/warning systems are activated or if a disaster happens without warning. This means that evacuation assembly locations have been chosen and advertized, and transportation and/or evacuation routes have been pre-planned and cleared for movement by vehicles or people to follow to safety. Pre-determined safe sites where displaced populations can be cared for should have sanitation facilities, be stocked with water, food, and medical supplies, plus cots and blankets, and have medical and other personnel to help people work though the trauma of a life-taking or life-threatening disaster. Finally, these sites should have communication equipment to send and receive updates on post-disaster conditions. Each location that has suffered through like disasters in the past should maintain search and rescue teams with the equipment they determine they will need on the basis of past experiences.

Clearly, many nations do not have fully developed effective programs for disaster planning and relief. Therefore, it is other nations, international organizations, and NGOs that do have experience and capabilities to cope with disasters that can and do come to the assistance both in caring for displaced citizens with water, food, shelter, medicines, and other necessities and with economic assistance with which to rebuild what was damaged or destroyed, but only when their offers of assistance are approved. These should be given without regard to international politics or national pride that could prevent assistance to help people in distress. These groups come to ravaged areas with personnel specialized in search and rescue operations and critical assistance (engineers, search dogs) and available equipment. In some cases, this might mean helicopters to bring the necessities of life to areas that cannot be reached by vehicles because of a ruptured infrastructure.

4.7 Afterword

In the past chapters, we have discussed strategies on how to work toward a stable global population and how to provide the essences of life for populations that increase from natural growth, demographic changes as people move from rural areas to urban centers, and from immigration. It is necessary as well to attract development projects that brings employment opportunities for growing population and in this way increase a municipality's tax base so that funds are available to support health and education programs for citizens. Development planning to reach this end is the theme of the following chapter.

References

1. Siegel FR(1996) Natural and anthropogenic hazards in development planning. Academic Press, San Diego, California and R. G. Landes Co., Georgetown, Texas, 300 p
2. Minoura K, Imamura F, Sugawara D, Kono Y, Iwashita T (2001) The 869 Jogan tsunami deposit and recurrence interval of large-scale tsunami on the Pacific coast of Northeast Japan. J Nat Dis Sci 23:83–88

Chapter 5
Development Planning: A Process to Protect People, Ecosystems, and Business Productivity/Longevity

5.1 Introduction

In the post-WWII period, much economic development planning was focused on getting the highest economic return from the funding allocated to development. This was the path followed by some government leaders and their advisory councils that were driven by achieving short-term economic success. They did not understand or understood but ignored the long-term environmental ramifications for their citizens, their municipalities, and their nations. Today, many nations make environmental protection a priority in their development planning programs. Some countries such as China have only recently begun to respond to citizens protests, about bad air (smog, power plant, smelter, vehicle, and industrial/manufacturing emissions), contaminated water and lack of water (industrial effluents, river diversions), and degraded soils and soil productivity (acid rain, lack of water for irrigation). Whether the responses are legislated, enforced, and effective is in question. This is a valid point because China, a country that in 2012, used one-half of the global tonnage of coal and that has plans to build 450 coal-burning power plants in coming years. The increased combustion of coal would intensify air pollution because of the emission of toxic metals (e.g., mercury), and sulfur dioxide gas (SO_2) that reacts in the atmosphere to make acid rain more acidic. The emission of CO_2 from these additional coal-fired power plants would be an added stimulus to global warming/climate change.

5.2 The Planning Process

Modern development planners have learned from the errors of the past and begin their assignments with four questions to be answered. First, what businesses exist in the locality where we want to attract investment? Here, we are seeking to achieve

© The Author(s) 2015

F.R. Siegel, *Countering 21st Century Social-Environmental Threats to Growing Global Populations*, SpringerBriefs in Environmental Science, DOI 10.1007/978-3-319-09686-5_5

mixed and balanced projects so that a downturn in one sector will not greatly slow down an economy but will allow a population to move forward. To this end, we may seek a cornerstone development venture that will attract others enterprises. This can be considered comprehensive planning and may take time to achieve. Second, what has to be done to attract development projects? Here, we assess human and economic resources, infrastructure status, and proximity to natural, educational, services, and other support resources. Third, what do we have to work with in terms of human and economic resources? This will determine whether there is a cadre of professionals that can, with the funding allocated, work out solutions to problems that might otherwise deter investors from locating a project at a proposed site. Fourth, what else has to be done in a practical approach to negotiate bringing sector investment to an otherwise desirable location? Such negotiations are generally about financial matters to be settled between government and investor negotiators [1, 2].

Whether a project is small or large, whether it is residential, commercial, agricultural, or industrial, whether it is government or investor supported, its development must be regulated to assure the security of citizens, the preservation of ecosystems, and the long-term viability of the planned venture. This means that industrialists and agriculturalists, for example, must institute controls that insure physical security, maintain clean air emissions, guarantee safe water effluents, and preserve fertile soil in the environment they occupy [3]. In addition, good planning requires the prudent use of natural resources, perhaps by allocation, so that they will be accessible to expanded populations in the future.

5.3 Goals of Development Planning

Development planning serves several purposes. First is a utilitarian aim. Here, the political and economic aims are to bring investment to a location and through this investment create employment opportunities for citizens. For example, in an industrial venture, this will be initially in a construction stage, followed by jobs in the operations of a project, and jobs in the support businesses that serve the main enterprise. This is a great social benefit. Taxed income from business profits after expenses and from employees wages will contribute to a locality's tax base. The optimal use of taxes in the delivery of a locality's essential services (e.g., water treatment plants) and social services (e.g., education, health care) is the desired outcome of careful, thorough planning, and the basis for a healthy, stable society.

Second is a security purpose. Here, the goals are to protect human populations from chemical, biological, and physical threats in air, water, and soils that could originate from development and be a risk to citizens health and well-being. Similarly, ecosystem habitats and natural resources have to be preserved as sustainable biomes protected from intrusions by pollutants and shielded where possible from natural hazards. This will allow ecosystems to continue to provide many basic needs for existing populations and for future generations.

Third is a safeguard goal. The aim here is to protect enterprises that bring employment to a community, better the quality of life for workers and their families, and improve its tax base. This means maintaining infrastructure that supports investors' ventures, and as discussed in the previous chapter, building according to an up-to-date code to protect against natural hazards and prepare plans to cope with them as needed. The specifications for a proposed project should require that the best technology be applied so as to forestall any harmful emissions or effluents from it. In addition, there should be plans for the secure disposal of liquid and/or solid wastes so that there are no environmental interactions that could degrade surface and subsurface ecosystems (i.e., surface water, soils, aquifers) near or at some distance from a functioning operation.

5.4 Attracting Development Projects

State governments and local municipalities are in competition with others to attract new ventures for the purposes cited above: (1) In line with project goals, development can create long-term employment opportunities for their citizens; (2) the government tax base will increase and can be applied to improve or provide social services (e.g., education, health care) and other government services (e.g., utilities, waste collection) for its population. To reach this end, a government has to determine what it has as selling points. This would include a site that has essential services and infrastructure (e.g., operating utilities that can be readily upgraded as needed, good transportation with road system, railroad, airport, and waterway). Other important selling points are the availability of an educated pool of human resources and a commitment to provide training according to the needs of an incoming project, access nearby to university expertise, and access to natural resources. Flexibility of governments in terms of negotiations on tax "holidays," willingness to build or contribute to infrastructure (e.g., access roads), and/or subsidies or grants can be a major factor in attracting investors to locate a business venture in a community.

With these and other positive factors as enticements, government agencies work to identify potential enterprises they can attract to their location. They next prioritize as targets in the quest for development those investment projects with needs that most closely match the "selling points." The next step is to invite a team from prospective investors in an enterprise to visit, judge, and rank the quality of the "selling points." This group will convey their assessment of the locale visit to a business committee that will decide whether to consider siting an enterprise at the location.

5.5 Stages in Development Planning

Small projects may be relatively simple to plan for. But as the size of an undertaking grows, development planning become more complex and complexities overlap to greater degrees with ever larger businesses. Whatever the size of a proposed venture, norms in its construction and operation have to be followed whether to building code requirements (or better) or to restrictions legislated to protect human well-being and viability of ecosystems, and vital natural resources.

Once a proposal for a venture is approved by a government development authority for further analysis in a permit granting procedure, the initial stage of an assessment of its viability in a location begins. The development authority assembles a project evaluation team. This includes experienced professionals (e.g., geologist, civil engineer, hydrologist, environmental scientist, biologist, chemist, waste management engineer, economist, statistician, sociologist, medical personnel, and other specialists as needed). The members of the team review the business proposal focusing initially on each specialty and then bring this expertise together to highlight the positive relations or negative aspects that have to be reconciled as a project evaluation advances.

This is followed by a thorough examination of recent and historical (in the past 10,000 years) natural hazards that have caused problems at the proposed location for the investor endeavor. These may or may not have been destructive, but if there was injury, death, damage, or destruction, it is incumbent on the evaluation team to discover why. Once completed, the team gives an estimate of the probability of potential problems from one or more natural hazard events recurring during the estimated life of the business. This is a basic step for a risk assessment in development planning. The team determines what was learned from past events that can be used to mitigate the impact of a similar future event and how this has been codified for implementation by planning commissions [4]. If not already in consideration, this information can be included to improve the proposal being evaluated.

An analysis of environmental problems caused by operating enterprises the same as that being reviewed is fundamental to thorough development planning. For example, problems can be from industrial and agricultural air, water (surface and aquifer), and soil pollution from emissions and effluents, and pollution from inadequately planned solid waste disposal that has damaged human and ecosystem health and productivity. Inadequate development planning in the past has caused or exacerbated hazard conditions (e.g., flooding, landslides, subsidence). These are happenings that planners study and can learn from to devise methods to counter the effects of an event. A project evaluation team has to assess these potential environmental problems, especially when there have been occurrences, for example, that have impacted active industrial or agricultural businesses that operate in similar geographic locations (e.g., climate, topography, geology). The team members have to determine if environmental problems have originated after a business started operation and when with respect to the start of its activity: short term and/or long term ones. If there have been intrusions into ecosystems, the team has to determine what has been done to mitigate or eliminate them. They ascertain how effective these counter measures have been for a specific issue or connected ones. If successful, these counter measures and any new technology can be included in a proposal for the development enterprise. The team calculates what the initial costs of installing environmental control equipment are and what annual maintenance costs are expected.

The evaluation team reviews compiled data and estimates the probability for potential natural hazard problems to recur during the expected life of an investment venture and beyond that time if the operation were to continue. This has been the case at many nuclear power facilities worldwide that have had their licenses extended beyond their decommission dates. The evaluation team issues directives for the full execution of protective/preventive measures that will assure, to a good degree, the security of the venture and its employees and their families. These directives include the following: (1) use safest geographic location for the project and its infrastructure elements; (2) apply revised building codes; (3) use best pollution capture/control technology and solid waste management; (4) use monitoring equipment and warning protocols; and (5) have evacuation/safe haven plans in place in the case that the methodology employed is not able to contain an event more energetic than what was prepared for. On the basis of a proposal with its cost estimates plus any added costs of protective and preventive measures and infrastructure that have to be installed if not already included, economists calculate an economic benefits versus costs relation for short- and long-term periods. Government decision makers consider the social and political benefits that can be derived for populations from the enterprise if it is approved by the investors against the costs of concessions they can make to investors to "bring home" a project.

In the penultimate stage of the planning process, the development authority and investors review the evaluation team report for a development scenario in order to establish its viability, especially with respect to the benefits/costs number for its estimated life. Also investors and government specialists look at the numbers but with the long-term sociopolitical benefits that can accrue to a population if a project locates in their jurisdiction. It is good public policy and sometimes critical to include citizens' groups (especially those with environmental concerns) in the review of the report and discussions that take place. The investors can accept the evaluation team report as it stands or requests additions such as subsidies that would give them some relief from the investment requirement. These may be tax exemptions or reduced taxes for a fixed period of time, government commitment to be responsible totally or in part for required infrastructure that will support the venture, training for future employees, and others. In addition, multinational investors will surely require that they be allowed to import equipment, export production, and repatriate profits freely or repatriate a portion of the profits if a portion is reinvested in the enterprise.

Negotiations follow that are meant to build up trust and confidence between the investors and the government. If conditions are met that satisfy both investors and jurisdiction officials, legal teams draw up contracts that are thoroughly reviewed by both parties. When acceptable, there is a document signing and the work to bring a business to operational status begins.

5.6 Afterword

Planning that attracts projects that provide jobs and thus improve the quality of life for workers in the business as well as workers in support industries are often based on the ready availability of natural resources to manufacture their products. In some cases, countries that supply the natural resources have limited the export of what factories in user countries need to function at full capacity. There can also be an embargo on the export of a commodity to user nations. A supplier country can do this in order to exert economic pressure that government officials calculate will enhance political influence over policies of countries housing the factories. The next chapter deals with this problem and how to counter it.

References

1. Waterson A (1969) Development planning: lessons of experience. Part 1. The development planning process. Jihn Hopkins Press, Baltimore, 365 p
2. Schilder D (1997) Strategic planning process: steps in developing strategic plans. Harvard Research Project. http://www.hfrp.org/publications-resources/browse-our-publications/strategic-planning-process-steps-in-developing-strategic-plans. Accessed 29 April 2013
3. Siegel FR (2008) Demands of expanding populations and development planning. Springer, Berlin, 228 p
4. Siegel FR (1996) Natural and Anthropogenic Hazards in Development Planning. Academic Press, San Diego, 300 p

Chapter 6
Exertion of Political Influence by Commodity-Based Economic Pressure: Control of Energy and Mineral Natural Resources

6.1 Introduction

Whether commodities are renewable (sustainable) or non-renewable, nations that lack them or do not have a sufficient domestic supply are in competition to obtain them to meet their domestic needs, for their agricultural and industrial interests and for national security. Much of the demand is from developed nations with industrial bases that help support their economies. However, increasing competition for commodities is coming from developing nations in Asia, South America, and Africa that are strengthening their industrial bases for goods they need or that they require for projects that provide or augment employment opportunities. This can help maintain social, political, and economic stability in nations at risk of popular protests because of unemployment. The principal commodities other than food, water, and shelter can be put in two categories: (1) energy sources and petro-chemicals and (2) earth materials (metals, minerals, rocks).

6.2 Energy

Energy sources that fueled economies and development in most nations during the beginning of the twenty-first century are extracted from the earth, finite and not renewable, and dominated by coal, oil, natural gas, and to some degree uranium. Wood converted to charcoal is an important energy source in the less developed world, especially in Africa and Asia. It can be thought of as not renewable because the rate at which it is being used is greater than the rate of growth of planted replacement saplings. These resources, with the exception of wood (charcoal), are moved across country between regions, and between continents as needs exist and economic capabilities allow. The supply of energy from these commodities can be disrupted temporarily or for extended periods by natural disasters (e.g., earthquake/

© The Author(s) 2015
F.R. Siegel, *Countering 21st Century Social-Environmental Threats to Growing Global Populations*, SpringerBriefs in Environmental Science,
DOI 10.1007/978-3-319-09686-5_6

tsunami, high category tropical storms), by embargo (e.g., political decision—OPEC, Gazprom), or by war, sabotage, and terrorism (e.g., damaging oil pipelines: Iraq, Nigeria, Egypt). Because of the possibility of disruption, nations accumulate stockpiles of commodities they consider essential for the continued functioning of their societies for given periods of time such as 3, 6 months, a year, or the period a government determines is necessary to resolve the problem of supply. The economic and political restraints on trade for these commodities are threefold: (1) which country is able to pay the price set for the commodity; (2) which country has a product or goods to trade for the energy source it imports; and (3) which country is willing to adhere to the international policies of the energy source country.

In response to using less of the non-renewable energy sources in order to alleviate production pressures, more efficiently use energy supplies, and decrease their dependency on outside sources, more nations are moving toward advancing their use of renewable energy sources such as hydroelectric, solar, geothermal, wind, tides, and waves. These energy sources are rather localized and can not supply power across oceans. However, hydroelectric power can be transmitted to bordering nations. Except for hydroelectric power, the contribution of renewable energy sources to the global energy supply is small. However, they are growing at a steady rate especially as research improves the efficiency of photovoltaic cells used in solar energy projects and improvements in the design of wind-driven turbines such as a pivoting head so that propellers can adapt to changes in wind direction to maximize electricity generation. The projected revenue growth from 2009 to 2019 for solar power was from $36.1 billion to $116.5 billion and for wind power from $63.5 billion–$114.5 billion [1]. Advances in technology have resulted in improved ways to store excess electrical energy from the renewable energy sources that can be drawn on if supplies from them falter (e.g., from less sun exposure or a drop in wind conditions). Clearly, the increases in the use of non-renewable energy sources, especially in developing countries, make a case for investing more funding in materials and engineering research to discover high-efficiency systems for the generation and transport of electricity. This can contribute to the energy needs for the growing global population as it moves toward stabilization by the turn of the century. It should be noted that hydroelectric dams are not truly a renewable source unless sediment that accumulates behind a dam over time is dredged out or flushed out periodically before its buildup reduces efficiency in the operation of the dam and its electricity output.

6.3 Earth Materials: Metals, Minerals, Rocks

6.3.1 Critical Metals—Strategic Resources

Earth materials that are extracted are not renewable, at least not within the context of humans' time on our planet. The question is whether the metals, ore minerals, and industrial minerals can continue to be extracted and processed at accessible costs for all nations and not mainly for the economically advantaged ones. For utilitarian purposes, countries classify earth materials as critical and/or strategic. Critical materials are those essential to industrial processes for the production of goods or components. Strategic materials are those essential to defense and national security needs and those required for economic development and public health and safety. The availability of these resources presents no problem if they are produced domestically in sufficient supply to meet a nation's demand. Otherwise, it is necessary to import them. Each country has its own lists of critical and strategic materials.

What makes these materials critical and/or strategic are first, their unique properties necessary for industrial production and for key defense applications. Second, there are no viable (economic) substitutes for them. Third, international supply chains for the metals are vulnerable to disruption, and there may be an import dependency that sets up the possibility of being cut off from them. For example, China produces 90 % of global supply of rare earth elements (REE). The over concentration of supply from a single country also presents a geopolitical risk of an interruption or cessation of delivery of the critical/strategic metals. An action such as this threatens the livelihoods of workers in industries that use critical commodities. China recently limited the amount of REE that could be exported to 20 % of its production, an amount that is far less than is needed to satisfy international demand, and potentially damaging to the manufacture of cell phones and other products in many countries.

6.3.2 Metals with Specialized Uses: Industry, National Security, Development, Public Health

Metals comprise most of the critical and/or strategic earth materials. They are found in a native state or bound in minerals. Many minerals bearing these metals and other essential industrial minerals have important functions for society. They are often found in granitic pegmatites and coarse-grained igneous rocks that crystallized slowly from magmas [2]. For example, the element lithium (Li) is a major low-cost component of rechargeable batteries and is used in specialized lubricants, and a Li compound is an anti-depressant and in the treatment of bipolar conditions treatment. Beryllium–copper (Be–Cu) alloys are in components of aerospace, automotive, and electronic devices. The metal rhodium (Rh), with almost all global production from South Africa, is required for the manufacture of automotive catalytic converters that change dangerous nitrogen gases from internal combustion engines into harmless forms. Capacitors based on the metal tantalum (Ta) are used in computers, smart phones, and for air bag and antilock braking systems in automobiles. "Blood" Ta is produced illegally by rogue militias using workers under inhumane conditions in the Democratic Republic of the Congo. Scientists are working on a technology to be able to identify Ta from this source and thus be able to alert responsible users not to buy it. The identification of "blood" diamonds from Sierra Leons and Angola has been in place for several years to alert responsible traders so they will not purchase them. Cesium (Cs), with most of its global production from Canada, is used to make high-pressure and high-temperature applications for petroleum/natural gas drilling. Its photoemissive properties make it an important component of solar voltaic cells. The element indium (In), with 80 % of its production in China, is essential in the production of LED lamps and is starting to be used in hyperefficient solar panels. An extended listing is given below.

The list is long and includes those important as ferrous (iron) alloys, (chromium [Cr], manganese [Mn], titanium [Ti], and cobalt [Co]), as industrial metals (iron [Fe], copper [Cu], lead [Pb], zinc [Zn]), and as precious metals (gold [Au], silver [Ag], platinum [Pt]). There are metals categorized as those that will continue to be important in future sustainable technologies [3]. They include the following:

(1) metals in electrical and electronic equipment to monitor political and social behavior that causes negative environmental impacts such as electronic devices in satellites for surveillance (e.g., Ta, Indium [In], ruthenium [Ru], gallium [Ga], germanium [Ge], palladium [Pd]);
(2) metals essential to photovoltaic technologies that provide power efficiency during production and consumption phases (e.g., Ga, tellurium [Te], Ge, In);
(3) metals used in battery technology to replace obsolete systems and reduce environmental impacts (e.g., Co, Li, In, REE);
(4) metals essential as catalysts for automobile emission reduction (e.g., rhenium [Rh], Pt, Pd, REE).

Other critical and/or strategic metals given in various tabulations are niobium [Nb], tungsten [W,] aluminum [Al], Be, Cu, Zn, Pb, nickel [Ni], tin [Sn], uranium [U], yttrium [Yt], scandium [Sc], molybdenum [Mo], vanadium [V], Ta, Ga, Au, and Ag. As stated previously, each country categorizes its critical and/or strategic metals needs from those cited above and others. It is notable that several of these metals have multiple uses because of their unique properties.

6.3.3 Industrial Minerals and Rocks: Critical and Strategic

Industrial minerals and rocks essential to industry and agriculture can fall in a country's critical/strategic materials category as well if not available from a domestic source. Industrial minerals are used in the manufacture of everyday products such as glasses and porcelains. Some are critical to the most advanced electronic devices such as in high-voltage electrical insulators (microcline), oscillators (quartz), process equipment for semiconductor chips, solar cells, and to make tiles that are extremely high temperature resistant such as those that were used on the US space shuttles for reentry shields (high purity quartz), printed circuit boards (kaolin), and spodumene (rocket propellant). These minerals are often found in granite pegmatites [4]. Other critical/strategic minerals and rocks include diamonds (oil/natural gas well drilling bits), phosphate rocks (fertilizer), graphite (nuclear use), sulfur (chemical manufacture), limestone (calcium carbonate rock, 101 uses) [5], salt (chemical use, food preparation), and gypsum (in drywall board for construction).

6.4   Control of Critical/Strategic Materials—Potential Conflict Among Nations, Among Users

There is resource competition within nations and among nations for natural resources in addition to food and water, top soil-rich land and timber, and ore minerals, because of economic and national security/defense exigencies and because of perceived wealth and political influence they are thought to convey. As discussed in Chap. 1, populations are increasing globally and are projected to stabilize at more than 10 billion persons by the end of this century unless "crashes" cause a marked decline in population growth. These factors, complemented by economic advances, especially in developing countries, result in increased demand for scarce resources of critical/strategic materials to the point where for some, demand exceeds supply. Scarcity of these resources for reasons given in the following paragraph can instigate mismatched competition within or between countries. This can spark conflict in order to gain access to sought for resources or can result in cooperation between competing parties, the sensible option. Nations live and let live chiefly because of their mutual economic interdependence, their mutual interests and values, and because of defense alliances.

6.5   Causes of Supply Deficits

There are limited supplies of many critical/strategic materials because of various factors, some social, some economic, some physical, and some political, so that supply does not meet demand. A societal problem is the lack of attention to recycling that can return a major mass of many commodities into the global inventory (e.g., 90 % of aluminum is recycled) or where there is a more rapid depletion of available reserves caused by overuse or misuse of specific metals/minerals. Economically, there has been an emergence of newly financially advantaged consumers who draw on goods that use metals/minerals that are in short supply. The exploration to discover new deposits of resources in demand is restrained because of costs, technical limitations, political decisions, and environmental regulations.

In the political arena, countries may be tight-fisted with the output of their deposits of critical/strategic materials because they are building up domestic industries in order to create jobs and increase their tax base. They may ban or limit exports, set restrictive quotas on a limited export allowance, and impose high taxes on exporters. In addition, there may be an inequitable distribution to importing nations of the limited allowed exports. The latter scenario can happen when a major producer nation reduces access to critical/strategic materials because the interests of the nation that absolutely needs what another country can supply are generally incompatible. This is not morally acceptable because such actions negatively affect livelihoods in the user nation and the ability of workers there to care for their families. That these enactments do not occur with great frequency is the result of "realpolitik" because major supplier nations need to export these materials in order to generate funds to keep investing to grow their GDP and thus maintain and increase employment opportunities as well as to draw down large fiscal deficits.

6.6 Improve or Extend Critical/Strategic Materials Inventory

The global stock of critical/strategic materials can be improved by exploration to find reserves beyond those in active development. However, as noted previously, this approach has limitations because there may be technical and environmental restrictions on exploration and exploitation of new deposits that could be discovered and worked. Certainly, public and industry sectors education on efficient resource use, regulations that limit use, or technology advances that reduce use in a process can help extend the supply of natural resources. Less traditional exploration and exploitation of resources on the ocean floor to increase the strategic resources supply may be necessary to preserve a nation's industrial production. Japanese scientists are evaluating deposits of rare earth elements (REE) in oceanic clay sediments. REE are a special group of 15 metals with unique properties that make them critical and strategic to industrial and developing nations. Alternative sources are being sought in light of the Chinese edict cited earlier that only 20 % of its REE production would be available to global users. As a result, Chinese restrictions on export of these elements drove up the price and led to other countries reopening mines that had been closed because they could not compete price wise with REE production from China. In addition, some manufacturers that used China as a source for these elements found substitute materials for their products. This has driven down the price of REE. Production from other sources is now selling in competition with China that some consider as an unreliable supplier that can severely limit or cut off supplies because of a political decision. Critical and strategic metals in deep sea floor massive sulfide deposits will ultimately be mined, as will metals in deep sea manganese nodules. Supply and price will determine when this takes place. Finally, possible interruptions in the supply chain or actual problems of scarcity from mines highlight the fact that it is essential to support research that has the potential to produce viable and economically accessible alternatives to those critical/strategic earth materials in short supply.

6.7 Geopolitical Strategies to Access Critical Metals and Strategic Resources

Nations offer tangible incentives to gain access to the critical/strategic materials needs they are obliged to import from other nations. These are numerous and varied. They are often based on low-cost, long-term loans and credits, construction or renovation of infrastructure such as pipelines to carry oil or natural gas, extension of electrical grids, building roads and bridges, and modernizing rail systems, airports, and seaports for exporter nations. They may also offer to erect and sometimes staff and supply medical clinics, build schools, offer scholarships to universities, help to install communications systems, and contribute to other needs that improve citizens' quality of life. Where China has supported such programs in Africa, the projects have been critiqued because Chinese labor is imported for the jobs easing employment problems at home but doing little to alleviate high unemployment in host countries. The expansion of trade between producer and user nations is a discussion point as well. Thus far, bilateral aid agreements have not included mutual defense aspects although trade in arms can be a major topic in negotiations. The cited enticements have been used, for example, by China, India, and Japan in African and Asian countries as they vie for political influence to access their particular critical/strategic resource needs such as petroleum, natural gas, and iron ore in Nigeria, Sudan, and Kazakhstan, respectively.

If enticements do not induce a supplier country to allocate critical/strategic materials to a user country for reasons that may be political or economic, and there are no other sources or alternative materials that can be used, conflict can result. Negotiators work to resolve these conflicts.

6.8 Mediation Between Suppliers and Users for Access to Critical/Strategic Materials

6.8.1 Negotiator Role

Negotiators for both the supplier and the user must be knowledgeable and understanding of the political, economic, and social implications of an agreement they are charged with reaching. Negotiations should be in the hands of professionals because the negotiation implications are often beyond the experiences of politicians and their advisory entourages with personal interests and short-term goals. Negotiators evaluate the potential long-term economic, political, and social ramifications for both the supplier country and the user country if an agreement can not be reached. They emphasize that long-term planning is essential as part of the stepwise process to resolve trade problem and avoid the possibility of near-term conflict whether it be economic or in the worst scenario combative.

6.8.2 Example of Conflict that Has to Be Negotiated

Such might be the case in the intensifying bitter dispute initiated during the autumn 2012, between Japan, China, and Taiwan on the sovereignty of a group of small islands between Okinawa and Taiwan in the East China Sea called Senkakus in Japan and Diaoyu in China and Taiwan. Japan purchased the island group from a private Japanese owner, but China does not recognize the sale of the islands that were annexed by Japan in 1895. Sovereignty gives control of rich fishing grounds and possible oil and gas reserves in the exclusive economic zone that surrounds the islands. This raises the possibility of armed conflict to keep them or try to seize control of them. Negotiations have begun as a side issue of the United Nations General Assembly.

6.8.3 A Negotiation Process

Conflict resolution between parties at odds over a natural resource that one has but denies access or limits access to the other that needs the resource is under continuous discussion [6, 7]. Experts in conflict resolution advocate a stepwise approach for negotiators [6]. The first step in any negotiation process is to establish trust between antagonists by dialogue. As trust evolves, a second step follows that further builds trust between the parties: an exchange of technical information pertinent to the scarce natural resource(s). This is followed by further discussion that explores where the opposing entities have common long-term goals and interests that supercede those deemed incompatible at the start of the discussion. If the countries in further negotiation judge that their long-term goals and interests are mutually important, they have to decide whether export of critical materials and/or strategic resources can proceed and if so, what proposed allocations are acceptable to them. Once a negotiated agreement is locked in, a final step in the resolution process is the preparation and signing of documents with guarantees of compliance with agreed to terms.

6.8.4 Commodities Not Used for Economic Gain or Political Persuasion

In modern times of shortages or disruptions in supply, food is not used as an instrument for economic gain beyond the world price, neither is it used to exert pressure on the politics of a receiving nation in favor of the producing nation. When there are extreme shortages because of weather or because of natural disasters, food is donated to sustain the victims (e.g., Haiti, Indonesia, in 2010). The food wrappings show the name of a donor country or of organizations with the hope that this will generate good will in future relations with a recipient country. Health care is a commodity that is brought to segments of society gratis in areas of countries that need it most by organizations such as Doctors Without Borders. Health care can be an economic commodity as well. For example, Brazil has a shortage of doctors and brings in doctors from Cuba with the approval of the Cuban government to work in towns away from major urban centers. The doctors receive a good salary, but Cuba takes two-third of their salaries to recompense the state that supported their medical education.

Although food and water as commodities are not used to coerce political togetherness, environmental scarcity of renewable resources can that can be used to sustain life and for other purposes can initiate conflict that can sometimes become violent. The renewable natural resources include freshwater, cropland, rangeland, forests, fisheries, and other wildlife. Conflict can arise from question of access to and control of them and from who has decision-making powers about sharing them, allocating them, and managing them in terms of rate of use. As with the non-renewable resources, negotiations that highlight shared interests, build trust, and emphasize long-terms goals, and interests can end in a compatible negotiated agreement that obviates conflicts [8].

6.9 Afterword

Many problems associated with sustaining our Earth's growing populations, albeit at a decreasing rate of growth, are being dealt with by local, regional, and national governments. The sustenance as we learned includes water, food, shelter as basics, then health care, education, and employment and other facets of everyday life. There are global dangers that are intensifying and increasing that need international agreement to lessen their threats to great numbers of existing populations and could be more so for future generations. In the next chapter, we will discuss some of these major threats that are planetary in nature and need for "now" action worldwide to preserve societal stability in the world. These are global warming and consequent climate change, degradation of arable soil and its loss to encroachment, terrestrial and marine environmental pollution, and the continuing repair of the degraded ozone layer.

References

1. Pernick R, Wilder C, Gauntlett D, Winnie T (2010) Clean energy trends. Clean Edge Inc., Portland, 20 p
2. Van Linnen RI, Lichtervelde M, Cerny P (2012) Granite pegmatites as sources of strategic metals. Elements 8:275–280
3. Buchert M Schuler D Bleher D (2009) Critical metals for future sustainable technologies and their recycling potential. United Nations Environmental Programme and Oko-Institute, 81 p

4. Glover AS, Roers WZ, Barton JE (2012) Granite pegmatites: storehouse of industrial minerals. Elements 8:269–273
5. Siegel FR (1967) Properties and uses of carbonate rocks: chapter 9. In: Chilingar G, Fairbridge R (eds) Carbonate rocks (physical and chemical properties), vol 9B. Elsevier, Amsterdam, pp 343–393
6. United Nations Environmental Programme (UNEP) (2010) Environmental scarcity and conflict: guidance notes for practitioners. New York, 70 p
7. United States Institute for Peace (USIP) (2007) Natural resources, conflict, and conflict resolution. USIP Press, Washington, DC, 31 p
8. United Nations Environmental Programme (2012) Renewable resources and conflict. Toolkit and guidance for preventing and managing land and natural resources conflict. United Nations, New York 116 p

Chapter 7
Global Perils that Reduce the Earth's Capacity to Sustain and Safeguard Growing Populations—Tactics to Mitigate or Suppress Them

7.1 Introduction

There are global happenings that imperil the earth's capability to sustain today's populations and the populations that are projected to increase until at least the end of this twenty-first century. These are: (1) global warming and the consequent climate changes; (2) soil degradation by erosion, loss of fertility, and other factors; (3) pollution of air, water, and soil from human activities and natural sources; and (4) to some degree, degradation of the ozone layer and its reconstruction time frame.

7.2 Global Warming/Climate Change

Global warming/climate change is real and progressive and is based on long-term measurements and observations. Since the late nineteenth century to 2012, the earth's average land and ocean surface temperature has increased by about 0.85 °C (1.53 °F). Measurements have been made at thousands of locations on land and in the oceans using various instruments and bathythermographs, and during the past 35 years, data gathered by satellites. The warming trend has been a subject of increasing scientific investigation during the past 50–60 years by scientists who determined that the increase in the global temperature between 1951 and 2010 was 0.6–0.7 °C [1].

The warming is caused in great part by the increasing concentrations of greenhouse gases (e.g., carbon dioxide [CO_2], methane [CH_4], nitrous oxide [N_2O], ozone [O_3], fluorinated carbons [e.g., CFCs]), and water vapor in the atmosphere abetted by particles and soot from the combustion of fossil fuels (e.g., coal, wood), and by deforestation. The blanket of greenhouse gases prevents the energy of the sun that hits the surface to fully radiate heat back from the surface and escape into

© The Author(s) 2015
F.R. Siegel, *Countering 21st Century Social-Environmental Threats to Growing Global Populations*, SpringerBriefs in Environmental Science,
DOI 10.1007/978-3-319-09686-5_7

space. The result is that some heat reflects back to the earth surface causing global warming. The Intergovernmental Panel on Climate Change (IPCC) estimates that during the twenty-first century, the global surface temperature could rise by more than 1.5 °C (2.7 °F) compared to the 1850–1900 surface temperature if emissions are kept to a low-emission scenario [1]. Today this seems unlikely. With low emissions, models calculate that the surface temperature will rise by more than 2 °C (3.6 °F) but less than 4 °C (7.2 °F) unless major volcanic explosive eruptions send ash into the atmosphere that circles the planet and causes a short-term global cooling. For example, this happened after the Mount Pinatubo eruption in the Philippines in 1991 that resulted in a 1 °C cooling of the earth for more than a year, after which warming resumed. The warming will be greater in tropical and sub-tropical regions than in the mid-latitudes. Nonetheless, the thoughts persist about keeping global warming from exceeding 2 °C (3.6 °F), a temperature that international bodies of scientists feel the earth and its ecosystems can adapt to [2]. This would be by adaptation to climate changes and their effects on many sectors. The track record of nations accepting limitations to greenhouse gas emissions from power plants, cement production, other industries, and agriculture is dismal for most countries given the reports emanating from the November 2012 climate change conference in Doha that was focused on arresting global warming and the climate changes it is fueling. To limit the global warming to a temperature the world can adapt to means keeping warming to less than 2 °C (3.6 °F). A warming exceeding 2 °C could play havoc with populations worldwide, especially in coastal regions.

Global warming causes physical changes that have chemical and biological ramifications that affect the planet's ability to sustain the Earth's population and its security at 7.2 billion people in 2013 and more so at the end of the century when the global population is estimated to reach more than 10.3 billion people. The 2007 IPCC four part report presented the existing data on climate change with assessments of the future effects of global warming and the resulting climate changes on our living planet [2]. The discussion that follows is derived in great part from the Panel's report complemented by the 2013 Executive Summary of the fifth IPCC report that will be published in 2014. Research papers subsequent to the 2007 IPCC report carry new data that have modified and sharpened the report projections of evolving future consequences of global warming on environments, their contained ecosystems, and growing populations, especially in Africa, Asia, Latin America, and the Middle East.

The alarming effects of global warming/climate change are many and interrelated. They are physical, biological, and chemical. These include rising sea level, latitudinal shifts of agricultural zones, migration of life forms, warmer ocean surface waters, shifts in flow paths of ocean currents, greater numbers of natural hazards including extreme weather events, and the geographic spread of diseases by insects, rodents, and other vectors. These are discussed in the following paragraphs.

7.2.1 Rising Sea Level

During the past 100 years, sea level has risen about 30 cm (8″). The rise has been determined from thousands of measurements by tidal gauges worldwide that recorded high-tide and low-tide levels to establish an average sea level for a given year. These measurements have been complemented since the late 1970s by satellite measurements that give a yearly sea-level average. From 1901 to 2010, sea level rose 1.7 ± 0.3 mm/year. Satellite measurements show that during that time span, the average yearly rise from 1971 to 2010 was 2 mm/year and the average annual rate of rise from 1993 to 2010 increased to 3.2 ± 0.4 mm [1]. This latter increase indicates a rise in the rate of global warming that is most likely related to the strong industrial advances and greater use of fossil fuels and increasingly higher CO_2 emissions into the atmosphere from developing nations such as China, India, and Brazil.

The causes of sea-level rise resulted from global warming are the thermal expansion of warmer ocean water and the melting of ice from alpine glaciers, ice sheets (e.g., in the Arctic), and ice caps (e.g., in Greenland and Antarctica). This explains about 75 % of the observed mean sea-level rise. Calculations in a recent study indicated that more than 50 % of the 8″ sea-level rise from 1902 to 2007 was from glacial melt [3]. For example, since the 1970s, Arctic ice sheets receded by 10 % per decade, had less thickness and a shorter duration, with melting especially high in late summer. This is not the case in the Antarctic perhaps because the ozone layer thinning and hole allows more reflected surface heat to escape into space. Thermal expansion of ocean water and melting of glacial ice goes on as global warming continues. There are widely varying estimates of how much more sea level will rise by the end of the twenty-first century. The IPCC estimates that the rise could be 18–59 cm (7.1–23 in) [2] whereas other researchers predict that the rise could be 15–22 cm (5.9–8.7 in) [3]. The IPCC presents a conservative model that gives a minimum range of mean sea-level rise of 0.26–0.55 m (10.2–21.4 in) and a maximum range of 0.46–0.82 m (18–32.1 in) [1].

As sea-level rises, encroachment on land will continue to a lesser or greater degree depending on the amount of the rise. Storm surges will put more land and citizens at risk as winds and seawater reach farther onto land wreaking havoc in coastal communities and driving saltwater onto agricultural fields. Additionally, seawater can invade freshwater aquifers that extend to the outfall at the continental shelf slope when the pressure of seawater pushing into an aquifer exceeds the pressure of the aquifer waters moving seaward.

7.2.2 Warmer Ocean Surface Waters and Shifts in Ocean Currents

Ocean surface waters warmed 0.44 °C to a depth of 75 m from 1971 to 2010 [1]. The upper 700 m have warmed as well. The warming of ocean surface waters has many negative consequences. Evaporation of warmer waters releases more moisture into the atmosphere to feed tropical storms/hurricanes. This can increase the strength of the storms that can drive storm surges inland. These can injure or kill coastal inhabitants who do not evacuate impacted areas, damage or destroy buildings and infrastructure, and produce heavy, sustained rainfall that leads to inland flooding. As noted previously, storm surges can inundate agricultural fields with saltwater damaging crops. Freshwater from excessive rainfall that causes flooding can flow into marine marshes and estuaries diluting their natural conditions and temporarily disrupting ecosystems that are essential to the propagation of marine life. Warmer ocean surface/near-surface waters disrupt cooler water ecosystems with the result that sensitive cold water life forms migrate from warming to colder waters. This alters the marine food web in terms of plankton, algae, and prey/predator fish relations that are fundamental to its normal pathways. This affects local fishing that is especially important as a food source for coastal populations in less developed and developing countries and also disrupts commercial fisheries.

There is some question as to whether global warming can alter the circulation of the North Atlantic current system and affect weather conditions in Northern Europe and Northeastern North America. The current system moves warm surface waters to the north/northeast toward the Arctic releasing heat to the atmosphere. This helps moderate winter temperatures in the high-latitude Northern Hemisphere. As the heat is released and waters become colder and denser, the current sinks and flows to the south in the deep ocean toward the middle latitudes (equator). In addition to the warming of surface/near-surface waters in the North Atlantic current system, two factors could change the flow path of the system, possibly in a century: (1) the influx of larger and larger volumes of frigid glacial meltwater caused by global warming; and (2) more precipitation at high latitudes that results from more ocean surface evaporation as global warming progresses [4]. This would likely cause longer and colder winter conditions in northeastern North America and in Northern Europe. However, although the current system circulation is predicted to weaken during the twenty-first century, it is not likely to undergo abrupt changes that will affect the weather greatly in these Northern Hemisphere regions [1].

7.2.3 Organism Migration: Consequences for Human Populations

Global warming is altering temperature conditions in terrestrial and marine ecosystems that drive the migration of life forms that cannot tolerate the warming and have the ability to move to more hospitable cooler ecosystems. For terrestrial organisms, this means migration to higher elevations or to higher latitudes. For marine and freshwater organisms that cannot tolerate the warming, the migration is to cooler waters. Those life forms that can move but still tolerate the warming ecosystems remain until the warming impels them to migrate to cooler ecosystems as well. The immobile life forms die, and if limited in population to an intruded ecosystem, go to extinction. Then, there are life forms that have nowhere to migrate to for feeding and breeding such as the polar bear and the arctic fox that inhabit shrinking habitats. They may adapt or face extinction. Indeed, bioscientists estimate that 17–37 % of our planet's species may go to extinction as a result of the failure or inability to adapt to global warming [5]. Unique tree species that have their populations in ecosystems with well-defined and limited temperature and rainfall ranges face extinction from warming and changing rainfall patterns. These may be physically moved to ecosystems where they can thrive but this "assisted migration" requires careful study by botanists and biologists to determine whether a transplanted tree species will have an impact on the new host ecosystem.

A team of researchers emphasized that the effects of a transition to a higher temperature regime on organisms would be seen first in the tropics [6]. Three factors enter into the researchers reasoning. First, the tropics are homes to most of the world's biodiversity. Second, tropical species live in climates with little variation and exist near the limits of their physiological tolerances. Third, the functioning of these species is disrupted by relatively small climate changes. Thus, it will take less warming in a future climate to cause noticeable changes. These changes (adaptation, migration, or extinction) will happen sooner in the tropics than in other regions.

Rising sea level and increased wave force during storms have eroded shores and forced human coastal populations to migrate inland. Global warming has resulted in expanding deserts driving human habitation toward semiarid or more humid regions. Greater rainfall that often accompanies the warming has expanded flood plains causing populations to migrate away from zones subject to repeated flooding. Humans adapt and survive. The same may not be true for some other life forms.

The grand majority of scientists accept global warming as a reality. Nonetheless, there is a small minority that believes the warming to represent a natural variation and is not abetted by the steady accumulation of greenhouse gases released into the atmosphere by human activities. An extensive study examined more than 29,000 data series statistically for temperature-related changes in physical and biological systems (e.g., organism migration and ocean current shifts, respectively) from 1970 to 2004 on all continents and for most oceans [7]. These data series were mainly from Europe and North America, with significant input from North Central Asia followed by very limited data series from Latin America, Australia, and Africa. The data collected since 1970 show that changes in natural systems were in regions of observed temperature increases. The scientists reported that at the scale of a continent, these natural system changes cannot be explained by natural climate variations alone. The 2007 IPCC report had concluded that most of the observed increase in average global temperatures on all continents except Antarctica since the mid-twentieth century was likely the result of the measured increase in anthropogenic greenhouse gas concentrations in the atmosphere [2]. On the basis of these data, scientists concluded that anthropogenic-stimulated climate change is having a marked impact on physical and biological systems globally and on some continents [7]. It should be noted that the warming also affected animals that follow annual migration cycles. Some birds, insects, mammals, and fish are migrating earlier. This has disrupted feeding patterns, pollination, and other species interactions that can affect human activities such as agriculture and requires human adaptation to the changing ecosystem conditions to maintain ecosystem vitality. For example, aquaculture fisheries may have to change species to those that can tolerate warmer waters, or if not, shift aquaculture farming to colder climates where their preferred species can grow normally.

A follow-up on the organism migration theme analyzed the data in 100s of journal articles on the topic of terrestrial organisms that are migrating to higher elevations or to higher latitudes as their adaptation to the warming climate [5]. The 764 species (23 taxa represented) in the evaluation of latitude change included insects, birds, fish, mammals, algae, intertidal species, herptiles, and plants in Europe, North America, and Chile. The 1,367 species (31 taxa represented) in the assessment of organism latitudinal change included insects, fish, herptiles, birds, mammals, and plants from Europe, North America, Asia, and Marion Island. The researchers calculated that the migration of species to higher elevations had a median rate of 11 m per decade whereas the migration away from the equator to higher latitudes had a median rate of 16.9 km per decade. They found that the migration distances are greater where levels of global warming are the highest and that the average latitudinal shifts were generally sufficient to track temperature changes. This shows for the first time a direct link between temperature change and shifts in the migration ranges. In some cases, the range of shift rates represented time delays in species responses to warming, whereas in other cases, it represents a species physiological response to the warming, and to interacting forces of change with the warming.

7.2.4 *Increased Spread of Infectious and Other Diseases to Human Populations*

Vector-borne diseases expand their ranges to higher elevations and higher latitudes as a result of climate change. This extends areas with warming temperatures and additional moisture into environments that can host the vectors. These include mosquitos and other insects, birds, rodents, and other animals. Also bacteria or viruses can be transported with soil particles to areas by winds during high-energy extreme weather that may be related to climate change. Table 7.1 shows selected vectors and the diseases they can spread when they invade environments that were inhospitable to them before rising temperature and increased rainfall.

Diseases that have spread into environments where they were previously unknown have increased human exposure and infected populations that are often not prepared to deal with them. For example, mosquitos have extended their ranges to higher elevations that in the past were cooler, drier, and inhospitable to them but are now warmer and have more rainfall to attract them with the *Plasmodium species*

Table 7.1 Vectors and diseases they can spread when they enter a hospitable environment [8]

Mosquitos: vector for parasites (e.g., *Plasmodium sp.*) that cause malaria, dengue fever, yellow fever, west Nile virus, and Rift Valley fever
Tse-tse fly: vector for the protozoa *Trypanosoma brucei* that enters a host and causes encephalitis (sleeping sickness). The disease is found in 36 countries in sub-Saharan Africa. There are 300,000 new cases that result in 40,000 human deaths in eastern Africa annually
Reduvid bug: vector for the protozoa *Trypanosoma cruzi* that causes Chagas disease
Tick: vectored by deer and other animals where ticks intersect with human populations and cause Lyme disease
Rodents: vectors of fleas bearing the bacterium *Yersinia pestis* to human populations that they infect with the (bubonic) plague. Changes in temperatures and rainfall that result from global warming are expected to expand distribution of rodent populations globally and spread the disease
Birds: vectors for avian influenza. Drought that may be a result of global warming causes infected birds to drink next to healthy birds and infect them. Persons who handle the disease can then become infected
Ebola virus: Outbreaks may be related to unusual variations in rainfall/dry season patterns. These variations are likely a result of global warming/climate change
Water: vectors for bacterium *Vibrio cholerae* that causes cholera. The disease is highly temperature dependent. Increases in water temperature that are caused by global warming correlate directly with the occurrence of the disease
Water- and food-borne: vectored intestinal and internal parasites cause diarrhea diseases. Rising temperatures and greater precipitation caused by global warming increase parasite survival and the incidence of these diseases

parasite they carry. This has caused the spread of malaria. For example, the New Guinea highlands at elevations greater than 1,800 m did not have environmental conditions that would host mosquitos and support their breeding. Similar conditions existed in the highlands of Ethiopia, Ruanda, Burundi, Kenya, and Tanzania. The fact that these areas now have mosquitos that infect populations with the malaria parasite is surely the result of warmer temperatures and more rain and moisture. Water pools up and creates mosquito breeding sites and warmer temperatures accelerate breeding. Other diseases such as those listed in Table 7.1 will likely spread as warming and moisture conditions increase regionally and allow the extension of the range of vectors that carry diseases to infect humans.

As warming continues to spread into new regions during the coming decades and populations continue to grow, vectored diseases will infect more and more people. The larger and denser populations will be susceptible to the continued spread of disease. Many of the newly infected people will be where population growth is projected to be high for the next 40 or so years (e.g., Africa, Asia, Latin America). These are regions where health clinics may be lacking, or clinics that are staffed with experienced medical personnel but lack access to necessary medical supplies. This will continue until existing methodologies and therapies are made available to them or there are newly discovered pharmaceuticals (e.g., vaccines or medications) made available that will protect people from diseases or control and eliminate disease. This has essentially been the cases for smallpox and polio, although polio eradication has been stalled in some areas such as in parts of Pakistan where Islamic fundamentalists are preventing the application of the polio vaccine.

7.2.5 Effects of Global Warming/Climate Change in Agricultural Zones

As noted earlier, the average mean global temperature has risen 0.85 °C (1.53 °F) since the end of the nineteenth century. The warming shifts toward the poles in middle and higher latitudes are driven in grand part by CO_2 emissions from sources such as the combustion of fossil fuels and cement production possibly aided to some degree by natural oscillations. The results manifest themselves in interconnected ways, some positive and some negative. To begin with, and as already mentioned, there is a shift in the ranges of plant and animal (including insects) species toward the poles and to cooler habitats at higher elevations. This causes changes in agricultural production patterns.

The warming has a positive effect of extending the length of a growing season for some agricultural zones. For example, in temperate latitudes of North America, the growing season has lengthened by 19 days over a two-decade period. Earlier sowing can result in an earlier greening of crops in the spring and an earlier harvesting of crops. Extended growing seasons open the potential for double cropping. One negative consequence of warmer temperature on agriculture is that a greater diversity of weeds, fungi, and pests (insects, animals) migrate to higher latitudes and higher elevations, proliferate in the warmer, more moist conditions, and can invade and damage or carry disease to crops, and other vegetation (including trees) growing there. Another negative effect of a warmer climate on agriculture is that higher temperature can speed up crop maturation such as leaf unfolding, not allow time for full grain growth, and result in reduced crop yields. For example, corn and soybean yields in the USA fall by 17 % for each degree increase in growing season average temperature. This is not the case for all crops. For example, early studies showed that extreme temperatures were harmful to rice growth and cause a decline in crop yields. However, a study of data on rice agriculture in China from 1960 to 2008, a period of a reported increase in global warming of 1.4 °C in the country, showed that the warming did not increase high temperature stress in irrigated rice paddies but did reduce the low temperature stress on rice crops across China [9]. This allowed for increased growth, development, and grain yield of this globally important commodity.

The 2007 IPCC report stated that the effect on world food production of a moderate warming of 1–2 °C (1.8–3.6 °F) may be small because the temperature rise is not the same globally [2]. Globally, scientists project that the surface temperature is likely to rise by more than 1.5 °C (2.7 °F) by the end of the century relative to 1850–1900 [1]. The tropical and subtropical regions will have larger temperature increases than those in the mid-latitudes. Reduced agricultural output in one region will probably be balanced by increased production in another region. This may be satisfactory for the 2013 population of 7.2 billion people (about 1 billion suffering chronic malnutrition) but will this balance serve the nutritional needs of 8.6 billion people in 2035 or will more people be afflicted with malnutrition? And how about 9.7 billion humans in 2050 or more than 10.3 billion by 2100 if these population numbers are reached? This is a frightening scenario for future generations worldwide.

The report estimates that crop productivity will increase slightly in middle to high latitudes with 1–3 °C (1.8–5.4 °F) increases in local mean temperature depending on the specific crop planted. High-latitude-crop-producing countries such as Canada, Russia, and the Scandinavian nations will likely develop larger cropping zones as the warming moves poleward. On the negative side, decreased crop yields are forecast for seasonally dry lower latitudes and tropical regions even at a warming of 1–2 °C (1.8–3.6 °F). Globally, the estimate is that the potential food production will increase with a warming range of 1–3 °C (1.8–3.6 °F) but decrease if temperatures exceed the upper range. More regionally specific scientists reckon that North America (exclusive of Mexico) is likely to have increases in aggregate yields of rain-fed agriculture by 5–20 % during the first few decades of this twenty-first century because of moderate warming and more precipitation [2]. Rain-fed cropping succeeds or fails depending on precipitation. A revised model indicates that mean precipitation will probably decrease in many mid-latitude and dry sub-tropical zones whereas it is predicted to increase in many wet mid-latitude zones [1]. Northern Europe is also expected to benefit from a warming increase in higher latitudes by increased crop yields (and forest growth). Negative projections are that Southern Europe will have a decrease in agricultural production because of the warming shift to the north. In the scenario of a Europe with a decreasing or perhaps stable national population, food security is protected. However, East and Southeast Asia could see crop yields increase by up to 20 % by 2050, but the yields could decrease by 30 % in Central and South Asia. These are regions with developing nations that have growing populations and agricultural land lost to urbanization, factors that put food security at risk.

Warmer temperatures in Latin America with its growing populations are expected to result in a decrease in productivity of some important crops. In drier zones, for example, a continuing temperature increase through 2050 would diminish agricultural productivity because of loss of soil moisture, salination, and spread of desertification. Compounding the future agricultural problems in South America is that climate change has caused a glacial retreat in the tropical Andes of between 30 and 50 % since the 1970s. Glacial meltwaters feed the rivers that supply safe water and irrigation water. A lessening or loss of this source of water would affect tens of millions of people who live downstream of the meltwaters and depend on it for potable water and for irrigation of farmland. A higher rate of melt as warming increases could lead to an essentially complete melt in coming years [10]. Researchers blame the loss of glacial ice on the average temperature rise in the region of 0.7 °C (1.26 °F) over the past 70 years.

Also in the Southern Hemisphere, the IPCC forecasts that agricultural and forestry productivity in much of southern and eastern Australia and parts of eastern New Zealand could decrease by 2040 because warming will add to periods of drought [2]. Drought hit Australia hard during the past several years and has caused an 80 % drop in rice production. On a positive note, areas in eastern and southern New Zealand near major rivers are projected to have increased agricultural production because of warming that extends the growing season, more rainfall associated with the warming, and less frost.

Of all the regions where food security will be at risk because of decreased agricultural productivity and growing population, Africa is of most concern internationally. In Africa, climate change is predicted to cause a reduction in average rainfall. This means that less water would be available for irrigation. Another effect of less rainfall and warming is that desertification would likely spread. Increasing populations in southern, northern, and western African regions will suffer increased water stress within a decade even as their populations expand greatly. In the same time frame, crop yields from rain-fed agriculture could decrease by up to half (50 %) in some countries causing malnutrition and starvation in their populations unless donor countries send in food for needy populations. Add to this an increase in soil salination in arable areas that could shrink marginal arid and semiarid farming areas that could see their crop yields reduced, and agricultural zones with a shortening of their growing season [2]. This portents badly for African nations. The key to meeting comestible demands midst the negative constraints that arise from global warming/climate change is adaptation to evolving agricultural environments. Adaptation as a counter to warming and change in agricultural zones will be discussed in the final chapter of this book.

7.2.6 Increase in Frequency and Intensity (Force) of Extreme Weather Events

Global warming is thought by many of the physical and biological scientists worldwide to contribute significantly to the increase in the number, intensity, and duration of extreme weather events. They include storms with violent high winds, storm surges in coastal zones, and very heavy precipitation, flooding, long-term severe droughts, extended heat waves and fires in dry, hot vegetated/forested areas, and extended periods of frigid weather. The effects of several of these natural hazards are to cause death, injury, illness, and damage and destruction of human settlements.

7.2.6.1 High-Energy Storms

Global warming has caused a rise in sea level and expansion of warmed seawater, thus increasing the area of the oceans. The warming is responsible for increased evaporation from the ocean and hence the greater moisture released to the atmosphere. Tropical storms tracking across the oceans suck up the moisture and gain strength from the ocean heat. This results in storms moving clouds with more rain that may precipitate back into the seawater, or when landfall is made, precipitate on land where it can run into streams and rivers or seep into soils and perhaps access aquifers. As wind speeds driving the storms increase, they can develop into extreme weather events called hurricanes in the northwest Atlantic Ocean, typhoons in the Pacific Ocean, and monsoons in the Indian Ocean.

The storms enter the extreme weather event Category 1 when their sustained winds exceed 119 km/hr (74 mph) and are considered very dangerous. Category 2 storms have extremely dangerous sustained winds of 154–177 km/h (96–110 mph) as experienced in October 2012 when megastorm Sandy impacted the USA east coast near shore communities (e.g., especially in New Jersey, New York City boroughs, and Delaware, as well as farther inland). The results from Sandy were exactly those described in the Saffir-Simpson hurricane scale and more: roofs were torn off, poorly constructed framed homes were flattened, insufficiently anchored signs, beach boardwalks, and piers were destroyed, small boats in unprotected anchorage broke their moorings, trees were uprooted and snapped, downing power lines that led to near total power outages for millions of people that lasted many days for some and weeks for others. There were scattered shortages of potable water. The "more" included heavy rainfall with flooding, and storm surges that carried destructive masses of seawater inland. The waters flooded the New York City subway system and closed it down for several days.

These storms become more devastating as wind speeds increase to Category 3: 178–208 km/h (111–129 mph). They cause increasing catastrophic damage when sustained wind speeds reach Category 4 with 209–251 km/h (130–156 mph) and Category 5 with wind speeds at 252 km/h (157 mph) or higher. This class of an extreme weather event has increased globally in frequency and intensity during the past few decades. Extreme storms have disrupted societal order and ecosystem equilibrium with wind, with flooding caused by heavy precipitation and run off, and with storm-driven seawater surges onto land that damaged structures, near shore farmland, and disrupted ecosystems in fresh water/brackish water marshes and in head zones of estuaries. One positive result is that more water in a drainage basin means that there can be more water to recharge aquifers depending on the rate of seepage, inflow, and runoff.

Communities can prevent deaths and injury from hurricanes (typhoons, monsoons) by first establishing warning systems that can reach all the public when meteorological teams report on the track and force of an approaching storm system. Second, plans to cope with disasters, as already mentioned, should have established evacuation routes that may be necessary and assembly points and transportation for those who do not own vehicles. This planning includes establishing safe shelters such as schools that are stocked with water, food, and medical supplies where people can be cared for if there is a call for evacuation or that can be readily stocked with these essentials upon storm warning. These locations should have telephones set up for the evacuees. Next, search and rescue teams should be prepared to respond to emergencies in a storm hit area. If possible, evacuated areas should have security patrols to prevent looting. In 1999, a powerful monsoon killed 10,000 people in India. A lesson was learned. The Indian government established shelters for evacuees in preparation for a future event. In 2013, the Category 4 monsoon Phailin with winds greater than 130 mph and gusts to 167 mph impacted India. A warning to evacuate locations at risk was heeded by almost a million Indian citizens that fled to the shelters. The death toll was 14.

Tornados present another class of wind-driven extremely high-intensity storm. They are not likely influenced by global warming because although additional moisture in the atmosphere fuels them, it also deprives them of wind shear that is a cause of much destruction. Nonetheless, they merit consideration in this section. Tornados originate with thunderstorms as violently rotating columns of air that are a funnel-shaped downward extension of a cumulonimbus cloud in contact with the ground. They rotate counterclockwise in the Northern Hemisphere and clockwise in the Southern Hemisphere, and generally have internal wind speeds in the vortex of 177–320 km/h (110–200 mph) that are close to the internal wind speeds in Category 3 hurricanes. Tornados are commonly 76–304 m (250–1,000 ft) across but in rare instances have been measured at 3.2 km (2 mi) across. They travel a few miles along the ground with the vortex kicking up soil, lifting and dropping debris, tearing apart structures in their paths, downing utility lines, lifting livestock and vehicles, and moving them to down path locations before dissipating and dropping them. Tornados can develop suddenly but warning systems can alert populations to the threat so that people can move to safe places in homes such as the southwest corner of a basement in central USA, to storm cellars, or where structures are built on slabs, evacuate to basements of nearby public buildings. Megatornados may cause destruction along paths of dozens of miles. Tornados kill and maim people and can wipe out towns. They can occur worldwide but develop seasonally mainly in the US mid-western and southeastern states.

7.2.6.2 Heat Waves

Heat waves are extended periods of extremely hot weather that may be accompanied by high humidity. The northeast USA sets a temperature of 90 °F (32.2 °C) for 3 or more consecutive days to define a heat wave but each nation or region establishes its own norms. For example, Australia sets 35 °C (95 °F) or greater for 5 consecutive days or 3 consecutive days at 40 °C (104 °F) or greater to define heat wave conditions. In general, heat waves last from 3 to 5 days but can persist for weeks. Humidity is important because together with the temperature, it sets the heat index that is used to alert populations to a health-threatening weather extreme.

Heat waves cause uncomfortable, and for some people, intolerable weather conditions that can lead to illness and death especially for senior citizens, children less than 4 years old, asthmatics, the chronically ill, and overweight people because of the way their bodies react to extended periods of unusually high temperatures. Heat waves are most dangerous to public health when they engulf densely populated cities that themselves are heat islands. Here, high daytime temperatures do not cool down much after sunset because concrete, asphalt, and stone absorb heat during the day and release the heat at night. Heat waves in urban areas with heavy city traffic and nearby industrial zones support the formation of smog and create conditions that favor wildfires in nearby vegetated areas that can then contribute to a smog condition.

There is little dissent from the observations and measurements that indicate that global warming enhances the probability of extreme weather events such as heat waves far more than it promotes more moderate events. Indeed, histograms have been graphed that show clearly the increasing frequency and intensity of hot seasons in the Northern Hemisphere during the periods from 1951–1980, to 1981–1991, to 1991–2001, and to 2001–2011 [11]. Evaluators of the graphs attribute this to the continual and increasing release of greenhouse gases into the atmosphere that abet global warming. An already warm climate would be expected to have hotter temperature extremes that cause more intense heat waves with more frequently record-breaking high temperatures then would be the case without global warming. The climate would also be expected to have less intense cold temperature extremes and fewer record-breaking low temperatures than without global warming. Both of these scenarios have been observed and measured. It is virtually certain that increases in the frequency and length of dangerous heat wave temperatures [heat indices greater than 38 °C (100 °F) or how hot you feel] and of warm daily temperatures over much of the Earth's continental areas will occur at a global scale during the twenty-first century. At the other end of the temperature spectrum, there will be decreases in duration, frequency, and frigidity of cold spells.

During late June to early August 2003, heat waves in Europe caused at least 52,000 heat-related deaths with more than 14,000 in France, 9,700 in Italy, and 2,100 in Portugal. The European summer of 2003 was hottest in 500 years. In Germany, temperatures reached more than 40 °C (104 °F) and in England, the high temperature was 38 °C (100 °F). In the Southern Hemisphere, 2008 and 2009 heat waves in Australia bolstered wildfires and drought. In South Australia, temperatures during January 2008 were 35 °C (95 °F) for 15 consecutive days and later in March, temperatures exceeded 37 °C (99 °F) for 7 consecutive days. Seven years after the 2003 European heat wave, the heat wave that hit Eastern Europe in 2010 exceeded the temperature maximum and spacial extent of the previous hottest summer in 2003 [12]. It affected 640,000 km² (400,000 mi²) or over 50 % of Europe including Western and Central Russia. The 2010 heat waves broke the 500-year seasonal temperature records set in 2003, and together with fires and resulting smog contributed to the deaths of 56,000 people. A great number of these were from the Moscow region that for the first time experienced temperatures of 38 °C (100 °F). Crops were lost so that President Medvedev banned grain exports in order to preserve the Russian food supply. That two megaheat waves would occur in the same region in the decade is alarming. The scientists who studied the heat waves presented a statistical model that indicated that if global warming continued, the probability of such megaheat waves occurring will increase by a factor of 5 over the next 40 years (by 2050) and likely be exceeded during the following 50 years [12] (by ~2100) when the global populations could reach more than 10.3 billion people, if this number is actually reached.

In addition to sickness, death from hyperthermia and other heat-related causes, and crop failure, resulting from high temperatures and lingering heat waves, there are other heat-related impacts from extended periods of high temperatures. These include buckled roads, warped rail tracks, burst water pipes, and power transformers that detonate and cause fires and electricity outages. Capability to repair these problems is essential in disaster response planning.

US Weather Service meteorologists have studied the anatomy of heat waves and can give warning of the possible onset of a heat wave. They determine when high pressure in the atmosphere at 3,000–7,600 m (10,000–25,000 ft) strengthens and stalls over a region for a period of time. Their experience with heat waves indicates that under the high pressure, air moves toward the surface and acts as a blanket that holds the atmosphere in and traps the heat, not allowing it to rise. Without a rise, there is little or no convection of moisture and hence cloud formation and little chance of rain. This results in a buildup of heat at the surface that people sense as a heat wave. As such conditions develop, meteorologists alert public health officials and citizens of an impending, health-threatening heat wave and other hazards to which it can contribute such as a toxic smog buildup.

To cope with heat waves, public health professionals have to plan for enough cooling centers to accommodate the citizens that will need such refuge and that have back up compressors in the event of power outages. The centers should be stocked with drinks that keep people hydrated and should also have medical assistance available to those that need it. In cases when people choose not go to cooling centers and they do not have air-conditioning, municipalities should make electric fans and water available to them to the extent possible.

7.2.6.3 Droughts

Droughts are naturally occurring because of below average precipitation over time that results in a deficiency in a water supply. Droughts can last several months or years. Consequently, there are below average water levels in river and lakes, lower water tables in aquifers, and a decrease of soil moisture in farmland zones. The latter is a cumulative condition that hurts agricultural crop and livestock production, thereby threatening food security. Regions in Africa, Argentina, Australia, and China continued experiencing drought conditions in 2009 that have lasted for several years. In addition to the shortfalls in precipitation, droughts dry out vegetation and set the stage for bush fires that can rapidly evolve into wildfires such as commonly experienced in southeast Australia, southwest USA, and other global regions.

Many scientists attribute extended drought conditions in grand part to changing climates resulting from global warming. Indeed, the IPCC reported that since 1970, as global mean surface temperatures rose, precipitation declined in the tropics and subtropics and that southern Africa, the Sahel region in Africa, southern Asia, the Mediterranean region, and the US southwest are getting drier [2]. The report indicates that globally, droughts are expected to be more intense, longer lasting, and cover more of the Earth's land area as the twenty-first-century progresses. This possibility has been questioned by other scientists who suggest that the increase in global drought is overestimated because the simplified model of potential evaporation used responds only to changes in temperature. Thus, it does not include more realistic modeling that accounts for changes in available energy, humidity, and wind speed [13]. When these factors are added to calculations, the results suggest that there has been little change in drought since about 1950.

To combat short periods of drought (perhaps 3–6 months), governments can collect water in surface reservoirs during periods of precipitation and runoff and store it, or in some cases collect and store water in prepared underground sites. Cloud seeding has been used to try to bring on precipitation. In near coastal areas, desalination plants can help maintain a water supply and mitigate the effects of drought. Extended droughts may require water rationing and/or the importation of water to sustain human populations. Otherwise, there can be a mass migration of people from water poor areas to those that have a water supply able to sustain the additional population.

7.3 Ozone Layer Degradation—Reconstruction

The ozone layer is in the stratosphere 15–30 km above the earth. It has protected life on earth from exposure to dangerous ultraviolet radiation (UV) from the sun by filtering it so that there is less exposure at the earth's surface. There are three types of UV radiation, each with its unique wave length range, designated UV-A, UV-B, and UV-C. UV-C is the most dangerous but most is absorbed by oxygen and ozone in the stratosphere and does not reach the earth's surface. UV-A is absorbed less by ozone and scattered less so that some reaches the surface. UV-B is mostly absorbed by ozone and is partly scattered back into outer space, thus limiting how much reaches the surface. An ozone layer monitoring network was set up worldwide beginning in 1928 and completed in 1958 to study the ozone layer and changes that may occur. Research showed that in the 1960s to the 1970s, there was a depletionor thinning of the ozone layer. A 20–40 % rise in skin cancer in the human population since the 1970s was attributed to the rise of 10–20 % UV-B reaching the Earth's surface as a result of the thinning of the ozone layer [14]. Australia with a high exposure to sunlight has a high rate of skin cancer and the USA has one million new cases of skin cancer annually.

7.3.1 Cause of the Loss of Ozone

The depletion of ozone is the result of chlorofluorocarbons (CFCs) accumulating in the atmosphere and releasing chlorine into the stratosphere where chlorine atoms catalyzed the destruction of the ozone [15]. These chlorine-bearing chemical compounds were discovered to be efficient refrigerants in the 1930s and were first used in refrigerators. Subsequently, in the 1950s, CFCs were used as propellants in spray cans, the major source of released CFCs, as refrigerants in air-conditioning units, and as cleansing and drying agents in industrial operations. This led to great increases in CFCs released into the atmosphere, their access to the stratosphere, and the resulting gradual thinning of the ozone layer that allowed the increase in UV-B reaching the Earth's surface.

7.3.2 Effects of Ozone Layer Thinning: Global Strategy to Halt and Reverse It

UV-B can damage DNA in the skin and mucous membranes. This causes genetic mutations that can give rise to skin cancer such as basal cell carcinoma, squamous cell carcinoma, and malignant melanoma. The knowledge that the thinning of the ozone layer was the cause of the increasing global incidence of skin cancer, and the discovery of the Antarctic ozone hole in 1985 [16], brought 191 countries to convene at a conference in Montreal in 1987 to discuss how to combat the problem. The delegates agreed to ban the use and future manufacture of CFCs over a period of time, to use up existing stock, and to find a substitute propellant for the spray can and other industrial uses. The resulting Montreal Protocol was issued, has been updated 7 times, and CFC production was banned globally [17]. As a result of adherence to the treaty, chlorine in the stratosphere is stabilized by 2000 and is declining. At the same time, ozone in the stratosphere is increasing, suggesting that the layer is starting to repair itself. The reconstitution of the ozone layer to its status before 1980 and its initial major interaction with CFCs will take some time because CFCs residence time in the upper atmosphere may be 50 years or longer as they deplete releasing lesser masses of chlorine that degrades increasingly less of the ozone. Some models have the ozone layer regenerated by between 2045 and 2065.

It should be noted that in addition to skin cancer, extended exposure to this UV radiation brings on premature skin aging, cataracts and other eye problems, and likely weakens the functioning of the immune system. It may even damage the DNA of crops such as wheat, corn, and soybeans so that seeds collected from them might affect the following year's crops [18]. It may also affect the reproduction of plankton in the oceanic photic zone and disrupt the marine food chain. The gradual reconstitution of the ozone layer will reduce these threats. An added benefit to the halting of the manufacture and use of CFCs is that they are greenhouse gases so that their reduction and elimination from the atmosphere will help the efforts to slow down and ultimately arrest global warming.

7.4 Erosion, Loss of Fertility, and Other Causes of Soil Degradation

Food-producing farmland is lost to urban expansion, and biofuel production. Also lack of water and loss of nutrients reduce cropland productivity. One researcher reported that these and other factors, soil erosion being the principal one, plus weather and disease, will reduce acreage for food crops 8–20 % by 2050 [19]. The following paragraphs will discuss erosion and other processes that degrade soils and how to manage them to maintain soil productivity.

Soil degradation is mainly the result of two principal processes: erosion, a physical one, and loss of fertility, a biochemical one. Both are worldwide problems that diminish crop yields and quality (nutritional value) whether in developed, developing, or less developed regions. Erosion and nutrient loss are especially acute in Africa and Asia where real-world conditions, especially economic constraints, such as the cost of fertilizers and lack of accessibility to modern slice soil/seed machines, do not lend themselves to the best agricultural practices.

Erosion or the physical removal and loss of soil (from agricultural fields) by flowing water or winds can result from one or a combination of causes. Soils can be exposed to erosion because of vegetation removal (deforestation), overgrazing, and by agricultural practices that overly rupture soil surfaces and expose soils to removal and translocation. Soil erosion has already caused a 40 % decline in agricultural productivity in many regions [19]. Overexploitation of farmland without nutrient replenishment or insufficient replenishment from growing season to growing season causes a loss of soil fertility. Plants unable to take up sufficient essential nutrients from soils (and air and water) will suffer. If a soil lacks one of the essential nutrients, plants will soon wither and die (Table 7.2).

There are ways soil erosion can be mitigated and sufficient nutrient levels maintained. Experienced management can limit soil erosion in several ways [21]. The principal one, as noted in Chap. 3, is to change tilling and harvesting practices such as traditional plowing as is used in many areas of Africa, Asia, and South and Central America that breaks up a soil exposing large surface areas of raised soil to removal by flowing water and wind. This is easier said than done because the preferred slice and seed equipment that cuts and seeds a field at the same time without breaking the soil, is too costly for farmers in less developed and many developing countries. The cost of fuel and maintenance is also a long-term concern to those who would use the equipment if it were made available. One would hope that governments, international aid agencies, and NGOs could provide this equipment that can be used by multiple farmers in a region with subsidized operating expenses as needed. Soil erosion on gentle slopes can be minimized by plowing at right angle to the slope. On steeper slopes, short-walled terrace farming can reduce erosion when the terrace walls that hold soil and water are maintained and

Table 7.2 Nutrients essential for plant growth [20]

Nutrients from air and water		
Carbon (C), hydrogen (H), oxygen (O)		
Nutrients from soil (also from lime and man-made fertilizers)		
Primary nutrients	Secondary nutrients	Micronutrients
Nitrogen (N)	Calcium (Ca)	Boron (B)
Phosphorus (P)	Magnesium (Mg)	Chlorine (Cl)
Potassium (K)	Sulfur (S)	Copper (Cu)
		Iron (Fe)
		Manganese (Mn)
		Molybdenum (Mo)
		Zinc (Zn)
Primary = used in largest amounts by crops		
Secondary = required in much smaller amounts by crops than primary nutrients		
Micronutrients = required in even smaller amounts by crops		

immediately repaired when a breech occurs. The preparation of soils in the spring when plant growth is high favors soil stability and limits erosion. Also, when crops are selected that come in at different times during growing season (multispecies cropping), the vegetative cover limits soil erosion. As an added benefit, this averts an economic disaster for a farmer if a monocrop is sown and fails because of disease. These methods serve to reduce erosion and are complemented when farm fields are further protected from the elements by windbreaks of trees and shrubs planted along their peripheries.

Nutrient levels in agricultural crops must be maintained to support optimum yield and nutrition value. Sadly, 86 % of crop land worldwide suffers some degree of loss of fertility and hence their productivity [22]. Two principal factors affect nutrient availability in farmland soils. First is overexploitation or cropping season to season without replenishing soil nutrients. This results in smaller yield and lesser quality of produce. Second is salination or salt caking on roots so that crops cannot take up nutrients although they are available in soils. As discussed in Chap. 3, this problem can be resolved by regular flushing of the soils that dissolves salt crusting the plant roots. Agriculturalists can maintain soil fertility from season to season by reviewing chemical analyses of soils that indicate which nutrients need replenishing. Replenishment is done using chemical fertilizers targeted to supply the deficient nutrient(s). Chemical fertilizer is added in amounts calculated to condition optimal growth for a given crop. It should not be added in excess because runoff could harm receiving ecosystems. A general replacement of nutrients by organic farmers is accomplished by plowing crop residues and compost (as available) into a soil. Conventional farmers use this method as a complement to chemical fertilizer.

7.5 Pollution Threats from Air, Water, and Soil—Human Activities and Natural Sources

Toxic chemicals that invade earth environments put human populations and other life forms at risk through their pollution of the atmosphere, the hydrosphere, and soils. These threats will intensify because of growing African, Asian, and Latin American populations exposed to pollution and their increased demand for goods and services and as middle classes expand their numbers in these and other regions. This will require more raw materials, chemicals, and energy to process them, then manufacture products, and finally use them. Only if chemical wastes from factories are captured and controlled at sites prior to any release into our ecosystem can the risk they pose to life on earth be countered. This is easier stated than accomplished because of political pressures, governmental interests, and presumed economic constraints. Nonetheless, in some instances, when a chemical threat to humanity intensifies and economic interests are at risk, governments have joined to alleviate it and ultimately to eliminate it. This has been the global response, for example, to acid rain, ozone layer thinning, and Hg and other toxins in the atmosphere that precipitate with rain, access soil, and enter the food web through food crops.

7.5.1 Atmosphere Perils

Coal-burning power plants and coal combustion for other industrial energy needs are major sources of atmospheric pollutants that threaten human health and that of other life forms as well as food security for global populations. The role of CO_2 emissions that contribute to global warming and climate change effects on the earth's ecosystems now and in the future were discussed earlier in this chapter. Other toxic emissions from the combustion of coal where emission control equipment (chemical scrubbers and electrical precipitators) is inefficient or not in use (e.g., at sulfide ore smelters, and at electricity generating power plants) include SO_2 (sulfur dioxide), the precursor to sulfuric acid bearing rain, and Hg (mercury), a neurotoxin that is especially harmful to young children and pregnant women. Additional potentially toxic metals such as lead (Pb), arsenic (As), and cadmium (Cd), and very fine particles (2.5–10 μm in size) also release into the atmosphere and invade the earth's ecosystems worldwide. The use of the best available technology can greatly reduce toxic emissions and the harm they can do.

7.5.1.1 SO_2—Sulfur Dioxide

The use of chemical scrubbers in chimneys has reduced SO_2 emissions by 90 % in many industrialized countries. However, developing and less developed countries exert less control legislatively on the release of SO_2 to their atmospheres that leads to the generation of sulfuric acid dominated acid rain. In China, for example, acid rain has gravely damaged farmland soils in many areas. Together with the loss of productive farmland to urban expansion, infrastructure development, land for industrial and recreational complexes, and natural disasters (drought, flooding), acid rain damage of farmland have put China's food security at risk. Arable land in China has been in decline from 127.6 million hectares (ha) in 2001 to 121.72 million ha in 2008. This is an average yearly loss of more than 825,000 ha that has to cease, not slowdown, but stop completely. To feed the Chinese population, the minimum farmland necessary is 120 million ha for crop production until 2020 to be self-sufficient in grain production if farmland loss stops, production is steady, and the population stabilizes. Failing this, China will have to import more grain than it does now to guarantee food security for its citizens. In 2012, China imported 1.7 million tons of corn. This amount is expected to rise to 5.7 million tons during 2013, and to 15 million tons by 2014/2015. Add to this 10 million ha of farmland damaged by pollution and a recent loss of 12 million tons of grain contaminated by heavy metals in the soil, the future food security problem in China becomes of great concern to those assessing it. The Bank of America estimates that there is now less than the 120 million ha in production (thus the corn imports), and that by 2015, there may be a decline to 117 million ha of arable farmland. However, China's media outlets (e.g., Xinhua News Agency and China Daily) report that the nation's productive farmland has been stable from 2009 to 2012 at 121.6 million ha. Minister Zhang Ping of China's National Development and Reform Commission reported at a 2012 Farmland Conference that China has <4.7 million ha that can be considered reserve farmland, a fact that may meliorate farmland confiscation for the sake of development. The loss of arable farmland is the outcome of planning by the Chinese government for the "now" economic gains against the recommendations of its own agricultural, environmental, and other assessments and planning agencies. In order to stem the threat to food security from the country's millions of hectares of arable land that has been lost, China owns two million ha of farmland overseas (mainly in Africa) and is negotiating to lease and purchase another three million ha of farmland in the Ukraine.

7.5.1.2 Hg—Mercury

In addition to SO_2 as a toxic product, coal-burning power plants emit the neurotoxin mercury (Hg). Mercury in the global atmosphere invades ecosystems as a volatile and via precipitation. In the oceans, this toxic element bio-accumulates in fish. Species higher in the marine food web (e.g., tuna, swordfish, king mackerel) are larger and bio-accumulate ever higher concentrations as they feed on smaller prey that themselves contain mercury. Mercury also accumulates in rice, vegetables, and fruits close to or down wind from emitting sources. When Hg-contaminated foods are consumed by humans, the toxin bio-accumulates in the body and can ultimately cause brain damage. It is especially dangerous for young children and pregnant women.

In industrialized nations, emission control units where coal is combusted capture much of the Hg before emission. However, the mass of Hg in the atmosphere continues to increase globally because of unused, inefficient or absent emission control systems in China's growing coal-burning power plants, ore smelters, and other industries that emit greater masses of Hg than the amount by which other nations have reduced in their emissions. China has pledged to retrofit older power plant systems and require the installation of technologically advanced equipment in newly built facilities in order to reduce Hg and other potentially dangerous metals emissions, but Hg in the atmosphere continues to increase.

7.5.1.3 Pb—Lead

Local atmospheric threats to humans and earth ecosystems exist at villages and towns close to and downwind of smelters, battery factories, steel and chemical industries, and waste incineration sites where emission control and capture equipment is lacking, outdated and inefficient, or available but not used. The Xinhua News Agency, the People's Daily, the China Daily, and the China Digital Times have publicized the poisoning of children and adults by lead (Pb) and other heavy metals in many Chinese towns and villages. This was from inhalation of Pb contaminated air, eating food tainted with Pb, and the normal hand to mouth transfer of Pb by children playing on contaminated soil. The ingestion of Pb in these ways, and its bio-accumulation, caused sickness in children, impeded their learning, and affected their behavior.

Severe Pb poisoning has been reported in Hunan, Henan, Yunnan, and Shanxii provinces. In all, 9 of China's 31 provinces had 1,000s of workers, children, and villagers suffering from exposure to toxic levels of Pb from January 2009 to June 2011. This has sparked protests by an enraged public because of children suffering from Pb and other heavy metal pollutants. The response of Chinese government officials has been to close factories that reopened a short time thereafter, fire managers and indict some. Overall, however, there has been little change and Pb and other heavy metals continue to be emitted to local atmospheres. Although there are 230 environmental laws in China, they have been rarely enforced in the three decade rush to rapid economic growth. If China closed three quarters of the Pb-acid battery plants over two to three years, this would reduce Pb demand by 70 %. Alternately, the Chinese government could use their foreign reserves to absorb the cost of installing or upgrading the emission controls at these sites with Pb capture technology if officials want to preserve the health of their young and old citizens alike. Other developing nations have similar problems but less grave because of more stringent control on emissions.

7.5.1.4 Dioxins

Dioxins are omnipresent, persistent organic pollutants. Dioxins are present in soils and sediments but more than 90 % of human exposure is through ingestion of foods, especially meat, dairy products, fish, and shellfish in which dioxins bio-accumulate in tissue. The chemicals are highly toxic to humans and can cause miscarriages, damage the immune system, interfere with hormonal activity, cause developmental problems, and possibly cause cancer. Dioxins originate as byproducts of various industrial processes such as waste incineration especially at uncontrolled, inappropriate electronic waste-recycling operations. In addition, dioxins come from smelting of ferrous and non-ferrous metal ores, manufacture of some herbicides and pesticides, as well as from past illegal dumping of PCB-based industrial oils. Dioxins enter the food chain via emissions carried to earth by precipitation and as waterborne pollutants. Internationally, dioxins concentrations in comestibles are continually monitored by specialized laboratories to keep the level in foods below what the World Health Organization (WHO) sets as acceptable.

China is one of the largest emitters of dioxins in the world with about half of the amount as emissions to the atmosphere, but the level of human exposure appears to be lower than in other countries. However, at sites of electronics recycling and waste incineration, the dioxin levels were high in breast milk of women working at these facilities. The Chinese government responded to this health threat by setting standards for e-waste recycling and incineration plants and for other key industries where dioxins originate. To minimize the generation of dioxins, incineration of wastes from these sites should be at temperatures from 850 to 1,000 °C, depending on the waste components. The Chinese government is attempting to reduce dioxin intensity by 10 % in 2015 and by more thereafter to reduce the health risk to its citizens [23].

7.5.1.5 Smog

Smog is a health hazard. It is a noxious mixture of gases comprised of various chemicals and fine-size particles (<2.5 μm) that accumulate but do not disperse in the near-surface atmosphere. Incidences of severe smog are increasingly affecting more urban centers that have growing populations worldwide. The major chemical in smog is ground-level ozone. Ozone forms when nitrogen oxide, volatile organic compounds (VOCs), and particles are exposed to sunlight that catalyzes photochemical reactions. The raw material for producing smog can be attributed to greater numbers of vehicles emitting exhaust fumes as a main source, to emissions from power plants and industries that service the growing populations and that are close to urban centers, and to the domestic burning of coke and soft coal (in China) for heating during cold weather and for cooking.

Dangerous smog commonly appears as a brown haze and takes several hours to form as component chemicals and fine-size particles accumulate in the near-surface air. Smog worsens and hangs over population centers when there is a thermal inversion so that denser cold air near the ground cannot rise through the warm air above it. Smog events often occur in valley sites where there can be a shielding from wind that could disperse the noxious matter. However, smog can also accumulate and blanket a lowland area surrounded by highlands, or even in lowlands when winds are negligible. Thus, meteorological conditions and topography have a great influence on smog formation and the length of time it lingers in an area. The accumulation of smog is especially efficient on hot summer days. Nonetheless, some of the worst smog conditions that filled hospitals with respiratory illnesses and cardiac problems have occurred in late autumn and in mid-winter. Cities with high populations and that are densely populated such as Mexico City, Los Angeles, London, Teheran, and Beijing periodically suffer from smog that is especially harmful to children and people with a history of respiratory and cardiac problems. Smog can kill as evidenced by 4,000 deaths from respiratory and cardiac stress in London during December 4–7, 1952, and 8,000 more deaths during the following weeks and months from its effects.

Many national agencies and municipalities require the monitoring of air quality to determine an alert level for smog. In the USA, this level is 84 parts per billion of ground-level ozone. When there is a smog alert, police and fire departments as well as radio and television announcements warn citizens to remain indoors and to delay intense outdoor activities until the alert is canceled. Some localities limit the number of automobile and trucks that enter cities when smog conditions exist. Governments may require that factories shut down until the smog threat dissipates. These are precisely the steps taken by the Chinese government in Beijing during January 2013. Beijing is a city with a metropolitan population of 20 million citizens encircled by a belt of coal-fired industries, including power plants that emit toxins and particles into the atmosphere. This, together with exhaust emissions from more than 5 million vehicles and emissions from many Beijing households that use soft coal for heating during the winter, add to the Beijing atmosphere smog load. The convergence of the introduction of emitted toxins into the Beijing atmosphere, little air movement, and a rare temperature inversion for a prolonged period brought on the smog problem. Air quality indices above a value of 300 are considered dangerous to human health. Measurements in Beijing reached 517, a value categorized as beyond hazardous. Measurements of particles in the noxious chemical mix in Beijing gave a maximum reading of 993 $\mu g/m^3$ of air, a value almost 40 times greater than the WHO safe limit of 25 $\mu g/m^3$ of air. These particles are small enough to deeply penetrate the lungs and enter the blood stream causing respiratory infections, asthma, cardiovascular disease, and lung cancer. In a smog event in 2012, exposure to ultrafine particles helped cause 8,572 early deaths in Beijing, Shanghai, Guangzhou, and Xian. The January 2013 smog event covered a great swath of northern China and affected more than 30 major cities. Then, on October 20, 2013, a heavy smog covered Harbin, China, and surrounding provinces. This coincided with warm temperatures and little wind, straw burning by farmers, and the start up of Harbin's coal-powered municipal heating system. The airport, schools, and highways closed for 3 days as the 2.5-μm particulate matter in the air rose to 1,000 $\mu g/m^3$, 40 times the WHO safe level, and visibility dropped to 50 m.

The fog dissipated completely by October 28 as a cold front brought snow and sleet in a front that moved in from Russia. China is planning to move to natural gas to fuel power plants but is still very much dependent on coal and continues to build coal-burning electricity generating facilities.

Two things can be done to limit the onset and the effect of smog. First and favored to reduce the potential for a sickening or killer smog is the mandatory installation, use, and monitoring of equipment for the capture of gases and particle emissions from industries, and an enforced stringent emissions control on vehicular exhaust. In cities where soft coal is still burned for cooking and heating, municipalities can subsidize a change to cleaner cooking and heating fuels. Second and applicable when meteorologists warn of an imminent onset of harmful smog conditions (e.g., little air movement, a temperature inversion, and other conditions that favor the formation of smog) is to temporarily shut down factories and limit the use and access of vehicles in cities under a smog warning alert in order to stop industrial emissions and reduce exhaust from vehicular traffic. All countries have environmental problems from pollution including heavy metal (e.g., Pb) poisoning and suffer through resulting health problems. The environmental problems in China are far reaching [24]. They have given rise to protests by angry citizen in many regions that have made spawned headlines in the Chinese press and other media outlets. In some cases, the government has responded by temporarily shutting down sources of the pollution. However, in order to maintain a strong GDP, offenders have not been cited by the government and some that have been cited have restarted their operations. For this reason and available reports in Chinese newspapers, China has been used as a prime example in the pollution discussion.

7.5.2 Waterborne Dangers

Waterborne threats to humans, other life forms, and to ecosystems are connected to one or a combination of factors. These include acidity, toxins (inorganic, organic, pathogenic), and fertilizer, nutrient, and runoff of hormone medication. In addition, there are contaminated fish, disruption of fisheries by migration as a result of warming ecosystem conditions, and invasive species of aquatic vegetation and fish/shellfish. Finally, there is what is been called acidification of the oceans.

7.5.2.1 Acid Mine Drainage

In a previous section, and using China as an example, we discussed the damaging effect of acid rain on soils and their loss of productivity. Acid rain events are relatively short, but they are repetitive albeit irregular and their effects on soils are cumulative. In place of soils, we consider here the impact of acid mine drainage (AMD), dominated by sulfuric acid, on terrestrial water bodies, with the understanding that acid rain can also contribute to the problems caused by strong acid discharge into streams, rivers, lakes, and associated wetlands. AMD affects many countries having historic and current mining industries and is a long-lasting problem. Mines worked in Europe during the rule of the Roman Empire more than a millennium and half in the past still discharge acid drainage. AMD usually originates from abandoned coal mines, mines for sulfide mineral ores (e.g., of silver, copper, lead, zinc, and iron), and gold, as the minerals and rock interact with water from rain or melting snow that infiltrates into the mines and waters that flow through and out of them. AMD is added to as the waters interact with waste tailings from the mines. The sulfide ores react with oxygen in the atmosphere and moving waters to form the chemical sulfate that in turn reacts with water to generate sulfuric acid. The drainage acidity has pH values less than 5.0 and is often as low as 2.0–4.5, levels toxic to most aquatic life.

AMD discharge disrupts the aqueous ecosystems and wetlands by either reducing population numbers, killing life at different levels of acidity, or causing organisms to migrate to water zones where they can tolerate the ambient pH. For example, AMD reduces benthic fauna on which food fish feed and can also be fatal to juvenile fish (e.g., salmonid fry). Clams cannot easily migrate and cannot survive when acid inflow brings river water pH below 6.5. Fish can migrate as they sense a slowly increasing acidity in river water. Rainbow trout can tolerate water pH of 6.0 but not less, whereas brown trout are safe at 5.5 but not a lower pH, and brook trout and yellow perch can live in waters with pH of 5.0 but not in waters that are more acidic [25, 26]. Studies in Pennsylvania, US mining regions, found that trout populations are found farther downstream and away from sources of AMD. In some cases, there have been massive fish kills as the result of a sudden uncontrolled discharge of acidic drainage and toxic metals flushed from mines and their acid/metal wastes and collection ponds into streams and rivers during strong and sustained thunder storms. The acidity of invaded streams and rivers is reduced by several pH units in 20 min, so that organisms do not have time to migrate away from the killing AMD.

An important damaging impact on streams and rivers that is a result of the acid drainage forming process is that it also releases iron and other metals into solution that can precipitate (e.g., iron oxide) or go into solution (e.g., copper). The iron oxide precipitate can destroy habitats by coating sediments and channel beds. Dissolved metals in solution, especially copper, can be toxic to fish and other organisms. In general, streams invaded by AMD are poor in taxa richness and abundance often with low to moderate numbers of only a few organisms. Conversely, healthy, unpolluted streams generally support several species and moderate abundances of individuals. In addition, there are more species of insects and algae in streams free of acid drainage.

On the basis of experience and many scientific studies at existing mines and their associated ecosystems, and in laboratories, it is safe to state that a country that awards permits for large-scale surface mining for sulfide minerals, native gold, or coal, for example, without stringent controls and continuous monitoring of fluid discharge, puts nearby habitats and organisms they house at risk. There is a good chance that surface waters will be polluted by AMD and toxic metals. This threatens the sustainability of safe water resources, ecosystems, and the fish and other assets they contain that are important to human populations.

The mitigation or remediation of existing AMD impacted streams, rivers, lakes, and wetlands is a problem with limited solutions without huge economic investment. Short-term fixes are designated as active and involve adding a neutralizing agent directly. These are based on the chemical reaction that an acid (e.g., sulfuric acid) reacts with a base (e.g., calcium carbonate) to give a salt (calcium sulfate) and water, with the loss of CO_2. Long-term sustaining solutions involve raising the pH of AMD and reducing its metal contents using containment and/or treatment plants.

Active remediation requires an ongoing input of economic resources to maintain neutralizing processes that in most mine drainage sites includes the removal of potentially toxic metals from the fluid [27]. In the carbonate neutralization process, ground up limestone (calcium carbonate) is reacted with the AMD to raise the water pH. This is a similar type of reaction that goes on in the stomach of persons with heartburn when they take an antacid such as Tums (calcium carbonate, chemically equal to the mineral calcite that forms the rock limestone). The acid is attenuated when it reacts with the antacid. However, the precipitation of the salt calcium sulfate as calcium carbonate reacts with sulfuric acid, may coat the limestone and limit it neutralizing effect. To maintain a raised pH as AMD continues to flow requires that limestone be added continually. This is an unsustainable holding action, not a solution to the problem but rather a steady draw on economic resources. Similarly, calcium silicate (chemically the same as the mineral dawsonite in the antacid Rolaids), a residue of industrial operations, can be fed into AMD and neutralize it to the pH that allows it to be discharged into an ecosystem. This is also unsustainable for the economic reason just cited. Some potentially toxic metals dissolved in the AMD may precipitate if the water pH rises into the basic range (greater than pH of 7.0), but those that do not can be sequestered by ion exchange. However, this would add to the draw on economic resources.

Lime (calcium oxide) neutralization is used to raise AMD pH. A slurry of lime is fed into a holding of AMD to bring the pH to ~ 9, a level at which most toxic metals precipitate. Iron and manganese do not but will precipitate if exposed to air (oxygen) in the system. The resulting metal sludge can be recycled in the process until the metals' concentration is saleable or is disposed of in a way that will not damage the environment.

A biological approach to raising AMD pH to safe levels and also remove potentially toxic metals is the subject of much interest. Sulfate-reducing bacteria are added to AMD holding ponds. They oxidize the sulfate-releasing bicarbonate that can neutralize acidity, and hydrogen sulfide that causes the precipitation of many toxic metals as sulfide minerals.

A passive treatment of AMD requires relatively little economic resource addition after the initial capital investment once the system is operational [28]. It raises the pH and reduces the heavy metals load of the drainage by directing the AMD into one or a sequential series of constructed wetlands designed according to the characteristics of the acid drainage from a specific source (e.g., pH, oxygen content, metals in solution, and their concentrations). In the passive process, AMD flows in (collection phase), is treated as it flows through one or a series of constructed wetlands (remediation phase), and finally discharges, clean, into an ecosystem. For example, AMD with iron that is readily oxidized (e.g., to rust) and manganese in solution, together with other potentially toxic metals that do not react to oxidization, can be directed first into an aerobic prepared wetland that may have a bed of triturated limestone (calcium carbonate). During this phase of the passive treatment, the iron and manganese precipitate and the limestone reacts with the acidity to reduce it and raise the pH. However, the precipitates coat the limestone slowing down the reaction and greatly reduce the efficiency of the process. The AMD can then be moved into an anaerobic constructed wetland with a bed of the triturated limestone and organic matter such as compost, sometimes with sulfate-reducing bacteria. In this phase, the limestone reacts with the AMD to raise the pH. The process is complemented by the anaerobic system that supplies sulfur from the organic matter or bacteria that reacts with any potentially toxic metals in solution to precipitate them as metal sulfide minerals. If the pH has risen to a safe value for the waterways (pH of 6.5), the drainage can be released into the environment. If the acidity of the waters is still too strong, the AMD can be moved into a pond with ground up limestone added so that the acid–base reaction can continue and bring the pH to an acceptable level for subsequent discharge.

7.5.2.2 Pollutants and Other Threats to Waters that Sustain Growing Populations

Groundwater not only provides potable water but is an important source of irrigation water for crops globally. Irrigated crops are major food sources for humans and food animals. If untreated effluents carrying inorganic and organic toxins discharge from industrial and manufacturing complexes, they can flow over and seep through soils and rock into aquifers, contaminating groundwater and also flow into surface waters contaminating them as well. This pollution can cause short- or long-term health problems for those who drink the water or consume the agricultural products grown with polluted water or fish/algae that feed and grow in a contaminated ecosystem. The solution to this problem is clear. Collect and treat toxic-bearing effluents before releasing them into the environment or recycle them into an industrial or manufacturing process. The technology is available to do this but is not used in many instances because of the economics to install it, use it, and maintain it, and the reduction of profits if it is installed, used, and maintained.

Inadvertent pollution of groundwater caused by human activity took place on the Indian subcontinent during the 1980s as the result of the green revolution that would improve food production. The green revolution required more well water for irrigation. Here, few wells existed before farmers dug and exploited 20,000 tube wells. High-volume pumping during the growing season lowered the water table seasonally exposing minerals in the aquifer rock to aeration and hence oxidation. One of these minerals is pyrite (iron sulfide) that contains arsenic substituting for part of the iron. The aeration and oxidation released arsenic into the aquifer water in the toxic arsenite form as aquifers recharged. Humans drank and cooked with the tainted water and ate produce grown with it. Over time they bio-accumulated the arsenic. This affected 200,000 people who became seriously ill and at risk for cancer and other illnesses. Millions were at risk in Bangladesh and India. The solution to the arsenic problem was to sequester the arsenite at the well heads or to collect and transport waters to a treatment facility before using the water. To prevent a similar incident, cuttings from a well-being drilled have to be examined and analyzed by geologists to assure that no such reaction as just noted will affect the good quality of the water source or make projects aware that treatment will be needed to clean the water before it can be used.

In some cases, irrigation waters from rivers carrying heavy metals in AMD have loaded these potentially toxic elements into farmland soil. There can then be a transference of this toxic metal through agricultural crops to human populations that consume the contaminated product. In Japan, an outbreak of the itai-itai disease (osteomalacia: bones become cartilage-like and death follows) was linked to AMD that carried cadmium in contaminated river waters that were used in rice cultivation, for cooking, and for drinking. Ingestion of cadmium through food (rice) and water and its bio-accumulation over time was the cause of the onset and progression of the itai-itai disease.

Only when there is a disaster and outcry from populations affected by unsafe waters for a period of time will some governments enact and/or enforce laws on waste control and the use of effluent cleansing systems. Such is the case at Lake Tai, China's third largest freshwater lake that supplied drinking water to more than 5 million people in the Wuxi City region [29]. Wuxi City has welcomed industrial development since the mid-1980s. It is the sixth largest industrial city and has 5,300 factories that employ half the city population. Industrial development proceeded with little care about environmental damage. Many factories released effluent with a toxic mix of pollutants into canals lakes, and rivers that flowed into Lake Tai. The contamination was so bad that from the late 1980s receiving waterways gradually changed colors, some milky white, others black. If a factory was closed because of egregious pollution, it was reopened after a minimum change in the effluent in order to maintain employment. By 1993, the entire lake suffered eutrophication (waters devoid of oxygen) and extraordinary algal growth stimulated by a nutrient inflow of excess nitrates and phosphates. When the masses of algae died, they decayed and decomposed in a process that consumed all the oxygen in the waters causing fish and other organisms to die. By 1994, Lake Tai was severely damaged with organic pollution of the water surface at more than 29 %. Seventy percent (70 %) of the drinking water supply was affected especially during May/April–September/ November. The rest of the year the algae disappears and the water is clean and clear. There was major investment in attempts to rid the lake of the algae problem such as moving water from the Yangtze River through the Wangyu River to the lake in 2002 to improve water quality but these were temporary and did not solve the problem.

The worst eutrophication occurred in May 2007 when the toxic mix of contaminated waters and sludge overwhelmed Lake Tai. Fish and other organisms in the lake died and the lake became rancid with toxic sludge. Mats of living and decaying algae in a bloom dozens of centimeter-thick covered the entire lake in June and emitted a rotten smell. The city government backed by the central government shut down or gave notice to 1,340 plants with those on notice having to clean up to meet effluent discharge standards set by the Chinese EPA by June 2008 or close down. In a major step to clean the lake, the central government allocated almost $20 billion in May 2008 for a 5-year plan to bring the lake back to an acceptable water quality. It is a major governmental error for any country to sacrifice environments and people in favor of economic growth.

Nutrients such as nitrates and phosphates and other agricultural chemicals from farmlands and animal husbandry run off into water bodies and cause algal blooms that harm life forms in rivers, estuaries, and oceans. As described above, this results from a eutrophic effect that comes on when algae die and use oxygen as they decompose, depriving fish of their oxygen needs, and bring on fish kills. In near shore coastal ecosystems, the input of excess nutrients stimulates the growth of dinoflagelates that use neuro-toxins to kill their prey. As dinoflagelates die, neurotoxins release into the seawater and cause massive fish kills. Problems such as these can be mitigated as emphasized previously if farmers use just enough of agricultural chemicals to grow and protect their crops so as to minimize any amounts in the runoff. The animal husbandry industries, especially commercial poultry operations with millions of chickens and beef producing feedlots with tens of thousands of cattle, and associated slaughterhouses are following environmental legislation where it is enforced by disposing of their nutrient-laden wastes so as to protect waterways and the ecosystems they contain and resources they supply to growing populations worldwide. In addition to aqueous ecosystem problems that can be caused by excess nutrients, there is the problem of improper disposal of medicines down kitchen drains or toilets. The medicines can be carried into rivers, for example, where they can alter the hormone activity in fish and give them bisexual identities. In many areas, medicines are collected by local governments biannually and are disposed of without a risk to the environment.

Another waterborne danger to human health that was discussed in a previous section is consumption of large food fish (e.g., swordfish, tuna) that bio-accumulated mercury, the neurotoxin especially hazardous for pregnant women and small children. Also, invasive species of fish and shellfish can thrive in new ecosystems and take over rivers killing or driving off native species and thus disrupting locally or regionally important productive fisheries. Municipalities use poisons to target the invaders and in some cases physical/electrical barriers to halt their migration.

7.5.2.3 Ocean Acidification

The process that leads to thinning of shells for some marine species or changes in growth patterns for marine vegetation is called ocean acidification. This is somewhat of misnomer. As described earlier in the book, acidity, and its counterpoint basicity (or alkalinity), is measured as a chemical value called pH on a scale that ranges from 0 to 14. The neutral value when a solution is neither acidic nor basic is 7. Values less than 7 are acidic and values greater than 7 are basic. Ocean pH values are basic averaging about 8.14, down from the 8.2 average at the beginning of the industrial revolution. There is a good correlation between the increase of CO_2 in the atmosphere since the industrial revolution and the decrease in ocean pH. Colder waters in the Antarctic, for example, hold more CO_2 and thus have a lower than the pH 8.14 average and are not acidic but rather less basic. Conversely, warmer water holds less CO_2 and become more basic.

Ocean water pH affects the saturation level of mineral components' marine life uses to precipitate shells that are comprised of calcium carbonate ($CaCO_3$) in the form of the minerals aragonite or calcite, minerals with the same composition but with different crystal structures. At a lower pH, the waters are less saturated with the mineral components and this makes it more difficult for marine organisms to pre-cipitate their shells. In addition, at a lower pH, there is a lesser bio-availability of essential trace elements such as Zn, Cd, and Fe for plants and animals such as corals (reefs), plankton, pteropods, and mollusks (clams, oysters). In seawaters with pH below the optimal pH value for calcium carbonate saturation, shells that precipitate are thinner as reported for the case of pteropods in Antarctic waters. Formed shells can partially dissolve in seawater well under saturated in shell mineral components. As CO_2 in the atmosphere continues to increase, the ocean acidification will worsen affecting critical ecosystem factors such the marine food web and the ocean's biodiversity. For example, coral reefs house a quarter (25 %) of marine biodiversity. In 2013, 60 % of the world's coral reefs were in waters with a lack of enough aragonite saturation and this affects coral animal's ability to precipitate their mineral structures. This in turn will affect the biodiverse population that depends on the coral reef ecosystem if reef formation/rejuvenation lessens in the future. This adds to the problem of evolving ecosystems that result from marine species migration caused by anthropogenic CO_2-stimulated global warming.

7.5.3 Chemical Threats to Human Populations from Soils

Soil chemistry is derived naturally from the rock from which the soil originated. In some cases, rock-forming minerals contain heavy metals that are inherited by a soil. In such a case, vegetables, fruits, and other commodities grown in the soil may take up and bio-accumulate one or more heavy metals. Consumers or users of tainted produce over an extended period of time can become severely sick. For example, in parts of Korea, rice, vegetables, and tobacco grown in a soil formed from black shale, a rock with many heavy metals, bio-accumulated the toxic metal cadmium. Over time, the ingestion of cadmium laden rice, other vegetables that bio-accu-mulated cadmium, and the inhalation of cigarette smoke from cadmium laced tobacco resulted in renal disease caused by the cadmium buildup in consumers' kidneys [30]. As noted in an earlier chapter, rocks that contain uranium emit the radioactive gas radon (Rn). Houses built over these rocks and/or soils formed from them have poor air exchange with the outdoors that can accumulate the gas. Inhabitants breathing the Rn overtime can develop lung cancer. Thus, pre-construction radon analysis of building sites is necessary so that a site can be avoided or venting systems can be built-in during construction.

Human activities can contaminate soils at their surfaces and from within a soil. In the first instance, for example, heavy metals can be added to soils, generally locally, by fallout of emissions from power plants, smelters, and other industrial sources if emission capture and control equipment is not functioning efficiently or is not being used. Previously, this chapter cited the problems from lead in locally grown produce in China, near battery factories, and smelters where fallout polluted soils with lead. Heavy metals in soils can be mobilized by acid rain and find their way into underlying aquifers and thus put those who use well water for drinking and cooking at risk of bio-accumulating them to toxic concentrations. Contaminated aquifer irrigation waters can transfer pollutants to soils and then soils to crops, making their use as food questionable when potentially toxic contents are detected.

Again, we stress that the application of manufactured chemical fertilizers to maintain soil fertility can cause problems if amounts used are in excess of what is necessary. Runoff of any excess nutrients into ecosystems can cause algal blooms that lead to eutrophication and fish kills. The use of herbicides, pesticides (insecticides), and fungicides to protect crops can kill some organisms in the soil that enhance soil productivity and as with excess fertilizers runoff into ecosystems to harm their productivity. Balanced and measured use of these agricultural chemicals can greatly minimize the negative effects on ecosystems from soil runoff.

Rainfall or meltwater runoff that contacts wastes and leaches inorganic and/or organic contaminants from disposal or landfill sites at the surface can contaminate soils and aquifers. This can be prevented by well-engineered sites. These include impermeable clay or plastic linings with built-in catch basins that capture polluted fluids that can be removed for treatment that neutralizes their harmful components before releasing waters into an ecosystem. Oil spills or leaks from breaches in pipelines can contaminate soil surfaces but these are generally localized and the company responsible for the pipeline responds to the need for a cleanup.

Soils can be contaminated from within by release of pollutants from buried waste disposal trenches of liquid toxic chemicals in barrels, buried chemical liquids storage tanks at industrial sites, and buried tanks of gasoline at dispensing stations. The release of these toxins is the result of corrosion of containers that breach, allowing the slow release of pollutants into a soil and thence to groundwater. There are thousands of these sites worldwide that were created before environmental awareness became important and before environmental laws were legislated that regulated disposal practices dangerous to the environment and populations and that included the requirement to regularly monitor soil contents. The worst of these have been or are being cleaned up but the process is slow and the task to clean up all the known leaking systems is overwhelming because of their number, the processes used, and the costs involved.

Finally, it should be noted that at some locations worldwide, soils are tainted from animal discharge of urine and feces. As explained in an earlier chapter, this can be stopped by using collection and treatment methods. In less developed and developing countries, where the cost of chemical fertilizer is beyond the economic reach of farmers, dried feces that bear pathogens have been used as a natural fertilizer. Through produce, pathogens can enter the food production process. Obviously, this is a danger to consumers.

There are several methods that can be used to deal with soils contaminated with potentially toxic inorganic and organic components. Most are very costly, may involve more than a single process, but are effective if the pollutant sources are eliminated. The method used depends upon the pollutant load. First is to excavate a soil, move it to a facility to clean it up before returning it to its original location. A second method is to combust the contaminated soil in furnaces at high temperature capturing emitted pollutants. A soil can then be replaced. A third method introduces bacteria into a soil in situ that consume or break down organic contaminants to harmless products. The fourth method is called phytoremediation and uses hearty, fast-growing plants that selectively take up one or more heavy metals. The plants are harvested after each growth cycle until a soil reaches its natural state. This will take several growing cycles. The harvested plants are either sold for their metal content or securely disposed of. The latter two methods that clean a contaminated soil in place are much less expensive. Although they may be slower, they can be very effective [7].

7.5.4 Biological/Health Dangers for Populations and Ecosystems

Biological threats to national, regional, and global ecosystems include the loss of habitat and hence essential biodiversity that disrupts food chains on land and in the oceans. This in turn leads to migration of organisms (e.g., predators) to new habitats where they satisfy their own feeding needs (e.g., new prey) at the expense of the natural residents of the invaded habitat. In some cases, the loss and changes in biodiversity can lead to extinction for some species. Human activity plays a major role in the biological changes by reducing organism habitats for human occupation and use, and by fueling global warming/climate change that triggers life form migration via emissions of CO_2 and other greenhouse gases into the atmosphere. Again, we note that the CO_2 and other emissions that disrupt the escape of the sun's heat from the earth's surface back into the atmosphere result in grand part because of two factors: (1) the rush of developing nations to industrialization, and (2) to meet the needs of growing populations in developing and less developed nations for housing, infrastructure, employment, goods, and services.

7.5.4.1 Surveillance, Eradication, Cures, Controls, and Maintenance

The biological health perils experienced worldwide are under constant surveillance by the WHO and regional groups such as the Pan American Health Organization and the US Center for Disease Control. Research globally is focused on the eventual eradication of life-threatening diseases. Only one human disease has been eradicated from our planet, and it was smallpox with no cases reported since 1977. The eradication was by vaccine.

Other diseases are near eradication. These include polio (by vaccination) with WHO reporting only 223 cases in 2012 (in Afghanistan, Pakistan, and Nigeria) with the expectation that eradication will be complete in 2014, and Guinea worm disease (by treating and making contaminated drinking water safe by purification, for example, with chemical tablets or by boiling). Measles could be eradicated by application of a vaccine where the disease is present but religion and political doctrine have not allowed the vaccination in some affected areas. Other diseases targeted by global health organization for eradication include malaria and dengue fever from parasites transmitted by mosquito bites, Chagas disease from parasites transmitted from animals by one of several biting insect species to humans, river blindness disease from parasites transmitted to humans by biting flies that can be kept at bay by a vector control program and a single annual dose of the drug ivermectin, Lymphatic Filariasis from a parasite transmitted by a mosquito bite that causes an infection with a disfiguring enlargement of limbs that can be cured by an effective drug combination if it were made available. Killer influenza viruses (e.g., avian flu, swine flu) are under constant research in order to develop vaccines against them and strains that mutate from them. Control of several diseases that are infectious human to human such as diphtheria, whooping cough, hepatitis B, tuberculosis, HIV, and others by vaccination or medication is a major focus of the WHO and other health control agencies. This will be discussed in some detail in the final chapter.

7.6 Afterword

As many citizens worldwide cope with growing populations and increasing population density, the quest for sufficient safe water and for untainted food puts them under continuous stress. They are under stress as well as they seek secure housing, good quality healthcare, education, employment opportunities, and personal and property security, and a promising future for their families. Their stress is compounded by the local, national, regional, and global threats to human sustainability that were discussed in this chapter. These stressors and strategies to relieve them are the subjects of the following chapter.

References

1. Intergovernmental Panel on Climate Change (2013) Fifth assessment report. In: Climate change 2013. The physical science basis. Summary for policy makers. Cambridge University Press, Cambridge, New York, 29 p http://www.ipcc.ch/report/ar5/wg1/. Accessed 18 March 2014

2. Intergovernmental Panel on Climate Change (2007) Climate change 2007 (as 4 part report). Part 1. The physical science basis (February 2007), Part 2. Impacts, adaptation and vulnerability (April 2007); Part 3. Mitigation of climate change (May 2007); Part 4. Synthesis report (November 2007). World Meteorological Association and United Nations Environmental Programme. http://ipcc.ch/publication_and_data/ar4/wg1/en/contents.html. Accessed 18 March 2014

3. Marzeion B, Jarosch AH, Hofer M (2012) Past and future sea level change from the surface mass balance of glaciers. The Cryosphere 6(6):1295–1322. doi:10.5194/tc-6-1295-2012

4. Curry R, Mauritzen S (2005) Dilution of the Northern Atlantic Ocean in recent decades. Science 308:1772–1774. doi:10.1126/science.1109477

5. Chen I-C, Hill JK, Ohlemuller R, Roy DB, Thomas CD (2011) Rapid range shifts of species associated with high levels of climate warming. Science 333:1024–1026. doi:10.1126/science.1206432

6. Mora C, Frazier AG, Longman RJ, Dacks RS, Walton MM, Tong EJ, Sanchez JJ, Kaiser LR, Stender YO, Anderson JM, Ambrosino CM, Fernandez-Silva I, Giuseff LM, Giambelluca TW (2013) The projected timing of climate departure from recent variability. Nature 502:183–187

7. Rosenzweig C et al (2008) Attributing physical and biological impacts to climate change. Nature 453:353–355. doi:10.1038/nature06937

8. Union of Concerned Scientists (2003) Early warning signs of global warming: spreading disease. Cambridge, Massachusetts. http://www.ucsusa.org/global_warming/science_and_impacts/impacts/early-warning-signs-of-global-4.html (2003 revision)

9. Sun W, Huang Y (2011) Global warming over the period 1961–2008 did not increase high-temperature stress but did reduce low-temperature stress in irrigated rice across China. Agric For Meteorol 151:1193–1201

10. Rabatel A et al (2013) Current state of glaciers in the tropical Andes: a multi-century perspective on glacial evolution and climate change. The Cryosphere 7:81–102. doi:10.5194/tc_7_81_2013

11. Hansen J, Sato M, Ruedy R (2012) Science briefs. NASA Goddard Institute for Space Studies. Columbia University, New York, 6 p

12. Barriopedro D, Fischer EM, Luterbacker J, Trigo RM, Garcia-Herrera R (2011) The hot summer of 2010: redrawing the temperature record map of Europe. Science 332 (6026):220–224. doi:10.1126/science.1201224

13. Sheffield J, Wood EF, Roderick MK (2012) Little change in global drought over the past 60 years. Nature 491:435–438. doi:10.1038/nature11575

14. Kane RP (1998) Ozone depletion, related UV changes and increased skin cancer incidence. Int J Climatol 18:457–472

15. Molina MJ, Rowland FS (1974) Stratosphere sink for chlorofluoromethanes: chlorine atomic catalyzed destruction of ozone. Nature 249:810–812

16. Farmin JC, Gardiner BG, Skanklin JD (1985) Large losses of total ozone in Antarctica reveal seasonal ClO_x/NO_x interaction. Nature 315:207–210

17. Montreal Protocol (1987) Full text can be view online http://unstats.un.org/pop/Documents/Doc0007.htm

18. Anderson JG, Wilmouth DM, Smith JB, Sayres DD (2012) UV dosage levels in summer: increased risk of ozone loss from convectively injected water vapor. Science 337 (6096):835–839. doi:10.1126/science.1222978

19. Ericksen PJ (2008) What is the vulnerability of a food system to global environmental change? Ecol Soc 13(2):1–18

20. Tucker MR (1999) Essential plant nutrients. North Carolina Department of Agriculture, Agronomic Division. http://www.adr.state.nc/agronomi/pdffiles/essnutr.pdf. Accessed 18 March 2014, 9 p

21. Blanco H-C, Lal R (2009/2010) Principles of soil conservation and management. Springer, Berlin, 617 p

22. UNEP Geo-2000 (1999) In: R. Clark (ed) Global environmental outlook 2000. United Nations Environmental Programme, Earthscan Publications Ltd., UK, pp 1–16

23. Zhao B, Zheng M, Jiang C (2011) Dioxin emissions and human exposure in China. A brief history of policies and research. Environ Health Perspect 11(3):A112–A113. doi:10.1289/ehp. 1103535

24. Hayes J (2008) Environmental problems in China: pollution, mercury, lead poisoning, cancer villages, and health problems. http://factsanddetails.com/china/cat10/sub66/item394.html. Accessed 18 March 2014 (updated July 2011)

25. Christensen JW (1991) Global science. Kendall/Hunt Publ. Co. Dubuque, Iowa, 296 p

26. Jennings SR, Newman DR, Blicker S (2008) Acid mine drainage and effects on fish health and ecology: a review. Reclamation Research Group Publication, Bozeman, 26 p

27. Johnson DB, Hallberg KB (2005) Acid mine drainage remediation options: a review. Sci Total Environ 338:3–14

28. Zipper C, Skousen J, Jage C (2011) Passive treatment of acid mine drainage. Virginia Cooperative Extension, Virginia Tech. Publication 460-133, Blacksburg, Virginia, 14 p

29. Guorui I, Hanfu H (2012) Long struggle for a cleaner Lake Tai. China Dialogue 14 Feb 2012

30. Kim KK, Thornton I (1993) Influence of uraniferous black shales on cadmium, molybdenum and selenium on soils and crop plants in the Deog-Pyoung area of Korea. Environ Geochem Health 15:119–133

Chapter 8
Stressors on Citizens and Ecosystems: Alleviation Tactics

8.1 Stress Defined

Stress on human populations can be defined by as a biological term. It is the psychological result of internal (physiological) reactions to external forces [1]. Stress evolves from a failure or inability of humans to adapt to change or respond at the right moment to emotional or physical threats whether actual or imagined. Contrasting responses may be to fight for what one judges to be right or to flee from a real or perceived threat to fight another day. Animals show the same reaction. Stress (psychological pressure) is caused by one or a combination of physical, chemical, biological, economic, or sociopolitical changes that imperil environmental vitality, sustainability, and stability. Stress on ecosystems results from a disruption that affects their ability to support the life and natural resources they contain. This is a consequence of an evolving or sudden, sometimes prolonged change in one or more factors such as temperature, precipitation or lack of it (floods, droughts), water salinity, pH, and influx of matter toxic ingested by organisms.

People suffer stress from working to meet the challenges of everyday living and trying to adapt to meet them. These include access to water, food, sanitation, housing, and health services as the most basic needs, and freedom, education, employment, personal security, property rights, and mobility as fundamental social needs. In ever more populated urban/suburban centers worldwide, citizens' mental strain is compounded by factors such as crowding, noise, traffic congestion, over-illumination, extremes of weather (e.g., high-energy storms, heat waves, droughts), or in the past three decades or more, the spreading menace of terrorism.

The stress process in humans starts with a state of alarm and adrenaline production, followed by a short-term resistance as a coping mechanism, and ends with mental exhaustion. Tension reveals itself by a wide range of emotional, behavioral (psychological), and biological-physical (physiological) symptoms. Common

© The Author(s) 2015
F.R. Siegel, *Countering 21st Century Social-Environmental Threats to Growing Global Populations*, SpringerBriefs in Environmental Science,
DOI 10.1007/978-3-319-09686-5_8

symptoms of stress in humans include irritability, muscular tension, inability to concentrate, headaches, and accelerated heart rate. Diseases such as ulcers, depression, and digestive system ills can be related to stress [2–4].

8.2 Stress on Humans

8.2.1 From Growing Populations

Stressors from expanding populations hit the very citizens that fueled the population growth in many affected nations (Chap. 1) and are felt by people directly or may have indirect effects (Table 8.1). Developed countries feel stress from global population expansion because of legal immigrants seeking employment and using social services and from illegal immigrants that do the same. The most severe stresses on growing populations to sustain healthy lives that have been emphasized in previous chapters are twofold. First is from not having access to a steady source of clean, safe water to drink, for cooking, for personal hygiene, and for sanitation, and second, from not having enough water for irrigation and animal husbandry to sustain food supplies. These are essentials that are missing or at subsistence levels for more than a billion people of about 7.2 billion on earth in 2014. In addition to constant causes of stress in society, there are those that may be temporary daily occurrences, especially in densely populated urban/suburban centers. These may include transportation problems such as overcrowded buses or subways, traffic delays driving to or from work, smog, noise, and odors.

8.2.2 From Economic Conditions

When there are "too many" people and too few jobs, there is direct psychological stress on those seeking employment and those dependent on them for the necessities of life. A naturally increasing population, added to by changing demographics as rural dwellers move to urban/suburban centers, and by skilled and unskilled legal and undocumented (illegal) immigrants fuels the "too many" factor. The lack of employment opportunities for citizens in many nations is exacerbated by employers that pay lesser wages with no benefits to immigrants than they would have to pay and provide to nationals. The unavailability of jobs is worsened further in industrial and manufacturing sectors by the increasing use of automation (e.g., robotics in the automobile industry). The creation of jobs thus becomes high priority for governments at all political levels.

Jobs are created by investors. Once a good employment level is reached, the stress caused by unemployment eases for those newly employed and their families and especially for young people in the 15–25-year-old range. To attract investment

Table 8.1 Direct and indirect stressors heightened by growing populations

Biological and physical
From less water and/or polluted water (chemically, biologically)
From less food, poor quality (less nutritious) food, tainted food
From bad air locally, regionally
From lack of sanitation
From encroachment on forests and agricultural terrain causing a loss of habitat in host ecosystems
Economic
Competition for natural resources based on supply and demand
Limited and focused investment, if any
Little employment opportunities or underemployment for capable cadre
Minimal acquisitive power beyond essentials, if that
No training or retraining programs
Poor infrastructure or decaying infrastructure no loans (funding) to improve or create infrastructure
Limited (but improving) communication networks
Small business loans may not be existent
Sociological
Increased population density and less green (open) space
Insufficient or poor quality housing
Limited availability of healthcare in general and local clinics in particular
Limited access to cooking/heating fuels
Limited access to electricity
Poor availability of child educational facilities, teachers, and tools
Political
Free vote, fixed vote, no vote
No elected government
Family, clan, tribal, ethnic, religious, regional, national loyalties
Lack of personal security
No habeas corpus
No free speech or right to assemble
No respect for human rights
Loss of property rights
Non-independent judicial system

These stressors put pressures on existing populations, productive ecosystems, and governments worldwide to start adapting to them immediately before reaching tipping points that could result in human disasters worldwide [1]

in large-scale projects, a government has to build a foundation to attract investors beyond tax relief and other benefits. First, it is important to have a stable government with a stable society unless an investment is made with political aims as well as economic gain such as South Korea, allowing factories to locate and function in North Korea. An investor requires an educated and skilled pool from which to select employees as well as a pool of reliable, willing to learn, unskilled workers. It is incumbent on governments to have training and re-training programs to fulfill this requirement. As emphasized in Chap. 5, an investor will want a good infrastructure in place (roads, railroads, bridges, airport, sea port, utilities), a reliable and secure communications system, and security for the public, including the employees and their families, and for facilities.

Governments that develop strong economies on the basis of export of commodities (e.g., industrial raw materials or finished products from them, agricultural, or energy) that keep their factories operating but when the economies in the export markets decline employees can suffer stress. The outcome is a slowing of production, furloughs, a loss of jobs, and a loss in government revenue, all of which damage the foundations of economic growth and can contribute to worker unrest. To counter this, it is essential to cultivate an economy with a domestic demand for the same goods that are exported. Failure to spur domestic demand for consumer products to make up for declines in exports will ultimately lead to economic decline. This requires employment at good living wages in order to increase domestic purchasing power and good quality factory or commodity output. Together, these should work to stimulate economic development and relieve societal and political stress. Whether national economies can continue to grow, nourish, care for, educate, and employ citizens as populations increase greatly in many less developed and developing countries are questions that are not frequently asked, and when asked, often go unanswered.

8.2.3 From Sociopolitical Strains

The two-time president of Peru, Fernando Belaunde Terry, was a respected practicing architect. He knew that a high density of population in the barriadas (neighborhoods) of poor people who came to Lima seeking employment, healthcare, and education for their children were causes of stress that fed unrest, violence, and crime within the barriadas and in the city itself. He reasoned that if new apartment building housing was built with the idea of opening up space so that people could escape the sense of confinement where they lived, that this would ease tension among the population and reduce some existing social problems. As president, he had apartment housing built with a ground level of columns, thus opening space for easy movement and with large green areas adjacent to or between the multi-family housing that allowed families and friends "place in the sun." Belaunde Terry's efforts were rewarded as violence and crime in the "new" neighborhoods and in the city decreased as the sense of being oppressed by too many people living in too little space was ameliorated significantly by using his architect's concept of open space. As global population grows, especially in developing and less developed nations, and as demographic changes have rural citizens moving to urban/suburban centers, the population density problem is likely to increase the "lack of breathing space" social tension. Addressing the problem by approaches such as that applied by Belaunde Terry and others with urban planning experience can improve the population experience and alleviate citizens' stress level.

Political action or inaction, decision or indecision, can cause stress and restlessness in a nation's population across all educational, social, and economic levels. A major dictum that has led to unrest and insurgency is to disallow free election of a national leader or a representative legislature, or to have no election at all. Societal stress heightens where there are prohibitions on free speech (critical of government officials or decisions), the right to assembly under the risk of arrest and incarceration, and a way to present grievances to a government. Without an independent justice system and provision for habeas corpus, there can be abuse of human rights, lack of personal security, and violation of property rights. Add to this, imposition of high taxes without improvement in public services (e.g., security, healthcare, education), infrastructure (e.g., in transportation, in delivery of utilities), and basic necessities of life, and societal stress and restiveness become protests for change. In historical times and in modern and contemporary times, a combination of a number of these factors has driven citizens to change governments, sometimes peacefully, sometimes violently, sometimes for the good, sometimes not.

Social strains among nations or within a nation can be the result of religious differences, ethnicity, tribalism, or clan loyalty. This can lead to oppression and undeclared or declared war. In 2013, the world continues to experience attacks on innocents by Islam's fundamentalists who demonize organized religions and different ways of life as they interpret the Koran. Within Islam, there is conflict between Sunni and Shia Arabs in some Middle East countries. There is internecine conflict between Fatah and Hamas Palestinians and between Lebanon's Christians/ Arabs and Hezbollah. Strains on populations from ethnicity and religious differences plus nationalism were the causes of the wars between Serbs, Croats, Bosnians, and Kosovoans in the states that were formed from the former Yugoslavia in the 1990s. Tension and hate between Hutus and Tutsis in Rwanda and Burundi where the Hutus comprise 90 and 85 % of the population, respectively, led to Hutus killing more than a half million Tutsis in 1993 in a tribalism driven genocide. Clan loyalties are evident in Iraq, Afghanistan, and Pakistan as conflicts continue in 2014 and tear at the fabrics in these nations that are in the process of defining their futures. Without tempering historically driven mistrust, intolerance, and hate born out of religion, ethnicity, and tribal/clan convictions, societal stress relief and peace among peoples will not be reached.

Another point of social tension is a consequence of the growing global population. This is emigration from developing and less developed countries with immigration to developed countries (as legal immigrants, seasonal and long-term guest workers, and illegal entrants). Professionals and others with skills needed by a country are welcome to immigrate whereas the less skilled and undereducated are allowed entry where there are not enough citizens or citizens unwilling to fill jobs (e.g., agricultural work, care for the elderly, garbage collection, other service positions). The expectation is that immigrants will learn the language, integrate into a nation's social structure, appreciate its culture, and become honorable and loyal naturalized citizens. The nation in turn has to accept the fact that the immigrants in their new citizens status will maintain their deeply rooted cultural principles while adhering to national doctrines and follow their religious identities while respecting other religions practiced in their adopted countries. Ideally, they will pass on these convictions to future generations as societal integration evolves and they meld with their fellow citizens in the general population. This was the case for Europeans that immigrated to the USA between 1840 and 1890 (mainly from Germany, Ireland, Sweden, Austro-Hungary, and Italy) together with emigres from Canada and Newfoundland, whether they were Catholic, Protestant, Jewish, or of another belief. The same is true in recent decades for Latin Americans and Asians. For example, these immigrants entered into the economy of the USA, availed themselves of opportunities to learn English, schooled their children, conformed to societal values, and became loyal naturalized citizens. They and their offspring joined native born sons to enter the military to defend their country when it was threatened and succeeding generations have done the same. They melded seamlessly with the general population while maintaining an unbreakable bond with their cultures and religious identities.

The "pass on" does not always take hold although most legal immigrants that have become naturalized citizens benefit from employment opportunities, healthcare, education, and other social/welfare benefits. There are clashes, often violent, that occur when deeply rooted differences exist between some alienated immigrant citizens and host populations such as have taken place in several western European countries (e.g., Germany, the United Kingdom, France, the Netherlands, Italy). There are murderous attacks on society by a small number of native born or naturalized persons such as detonating explosives on subway trains in London and exploding bombs at the Boston marathon. These are often driven by religious zealots or those with a loathing of a mode of government or a way of life. The rejectionists of a country that has nourished them, of its values and way of life, bring undue prejudice from the general populace on the vast majority of naturalized citizens of the same ethnicity or religious beliefs who follow the laws of their adopted country and the freedoms it offers. This is a major social stressor that has to be relieved by government security teams that gather intelligence and act forcefully on the basis of the intelligence to protect citizens from such murderous and injurious attacks.

8.3 Ecosystem Stressors: From Human Activities/from Nature

Ecosystems are stressed by human activities, by events that likely occur because of human activities, or by natural events. The result is a stress build up in citizens that depend on ecosystems for sustenance and natural resources.

8.3.1 Human Activities

8.3.1.1 Water Diversion

Water diversion projects can harm ecosystems in multiple ways if they are not continually well managed so as to prevent or greatly alleviate their impact on waterway environments and their juxtaposed wetland and terrestrial biomes. For example, as results of lessening water flow downstream, or diverting water flow completely, or causing changes in water chemistry and nutrient delivery, habitats are lost especially for fish, amphibians, and birds (along their migratory routes). Reduced water flow will cause silting of water channels and added pollution in some sections of diverted waterways. There will be a loss of biodiversity because of altered physical and chemical conditions in breeding grounds such as type of sediment transported and deposited, temperature changes in shallower water, and water chemistry changes in riparian habitats of rivers and streams immediately downstream. Where dams are built to control water flow, they can damage wetlands and also bar access of salmon to their natural upstream spawning habitats unless a managed system provides access. A result of these problems is that human populations suffer because of lost food and material resources such as unique sources of raw products for the development and preparation of medicines that can help humankind. Where water diversion projects are implemented at different locations along a river channel, habitats may be fragmented into expanses too small to support a natural biome. In the past, the physical diversion of rivers caused social, biological, and economic disasters that endangered the health and productivity of ecosystems.

A major disaster that originated from diversion of rivers away from a water body they fed was brought on by poor planning and management in the early 1960s by the USSR. When the former Soviet Union government diverted two rivers to provide irrigation for cotton and rice cultivation, the Aral Sea that was the fourth largest water body on earth and had been fed by river flow shrank and over time was reduced to less than 10 % of its former extent. The ecosystem suffered extreme stress with the loss of habitats as a saline desert took the place of the sea. Fisheries and canning industries that were the livelihood of local populations were destroyed, as was economically important seaside resorts' tourism. The less than 10 % of the sea that remained was very polluted and a threat to public health. After the disintegration of the Soviet Union, the nations bordering the sea, Kazakhstan and Uzbekistan, supported by the World Bank, began a project that is on the way to reclaiming the biodiversity and economic benefits of the North Aral Sea (on the Kazak side). The expectation is that reclamation of the South Aral Sea (on the Uzbeck side) will begin in the near future. Citizens in the region sensed that the climate changed because of the decimation of the sea.

In a contemporary project, China approved a 50-year plan in 2002 to divert water from the south where it is plentiful to the heavily industrialized north with high water use but low rainfall and where rivers are losing volume and flowing low.

The estimated cost of the three route project is US$74 billion and will divert water from the Yangtze River to the Yellow River and the Hai River. An eastern route that is partially operational moves polluted waters that can be used only by some industries until pollutants are removed. A central route, where industries and farmers had to reduce water use, is partially functioning and moves clean potable water from the upper reaches of the Han River, tributary of the Yangtze River, to Beijing and other major population centers. A western route that will draw from the Yangtze and Yellow Rivers is still in the planning stage.

This project is controversial. For example, it heightened social and economic tensions because it displaced hundreds of thousands of people including farmers who were compensated with lower grade farmland to grow rice, a crop they had no experience growing. More than 467 km^2 (46,700 ha or 116,750 acres) will be damaged upon completion of the project. Thus, the ecosystem farmland will drop out of food production and affect local food security. Water will be lost by evaporation and to pollution and the cost of clean water will be higher for the consumer causing economic anxiety. Critics also argue that diversion during dry seasons will lower source river water disrupting habitats and also be an impediment to ship transport along the major commercial route that is the Yangtze River. The Chinese government is working on solutions to many controversial points through its environmental scientists and engineers. Time will tell whether the project functions as planned or has failings that can be corrected without creating new problems.

8.3.1.2 Human Encroachment

Human encroachment invades grasslands, productive farmlands, and forested areas to open land for construction of housing for increasing populations, for infrastructure (e.g., roads, utilities) to support urban expansion, for commercial zones, and for manufacturing and industrial development. Encroachment is a worldwide problem that is stressing ecosystems and in some instances, such as in the Amazon, indigenous groups that live off the land and have had no contact with the outside world. This is more predominant in less developed and developing countries. This intrusion stresses and may destroy ecosystem habitats. Forested areas are leveled by slash and burn to provide land for money crops such as soybeans and for animal husbandry (e.g., in Brasil's Amazon), for food crops (e.g., in Kenya), and to harvest wood (e.g., in Borneo, Brasil). This encroachment onto natural habitats displaces life forms from their natural food/water sources and drives them where they may not be welcome (as invasive species) or where they cannot survive. In some cases, loss of habitat can lead to extinction of immobile life forms that are unique to an ecosystem unless they are protected by laws that are enforced, or if they are translocated to an ecosystem where conditions allow them to survive and flourish.

The stress on ecosystems from encroachment to accommodate urban growth driven by population increase can be alleviated in some localities by the use of vertical housing in urban centers with an established infrastructure that can be readily modified or minimally added to in order to service more people. As indicated earlier in this chapter, it is good planning to provide open space to counter social problems that could arise from high population density. Industrial parks have to be accommodated because they are a source of employment for growing populations with a well-educated cadre. Also, they provide revenue for governments. To this end, industries can be built with small footprints and with pollution control systems that minimize their impact on people and unspoiled land from emissions and/or effluents.

An illustrative example of the results of uncontrolled encroachment that has economic and social ramification can be found in Kenya [5]. Kenya's population is about 42 million people with an annual rate of population increase of 2.7 % (fertility = 4.9). At this rate, the population will double by 2040 and the food supply will have to increase accordingly. How and at what price is in question. In Kenya, the Mau forest land has giant trees that form a canopy with hardly a break and with a grand diversity of flora and fauna. Homesteaders came in with slash and burn in many large areas to open land for agriculture and infrastructure to support it with the excuse of needing land to cultivate food crops for the more than 1 million additional Kenyans born each year. Where there was a loss of canopy, scientists measured a $\sim$1–2 °C ($\sim$1.8–3.6 °F) rise in the uncovered land temperature over the past 3–4 decades. This stressed vegetation and food sources for wildlife as well as some predator–prey relations in the food cycle. Together with poaching, this caused a loss in biodiversity due to wildlife migration and for some species led to extinction. The attack on natural resources in Kenya since the late 1970s as a result of population pressure, land fragmentation, and poor farming practices that degraded natural habitats and caused a steady decline in biodiversity with the result that about 50 % of the country's wildlife population has been lost. Initially, areas with good potential for farming were invaded but since 2000, the arid and semiarid lands, excellent habitats for 80 % of Kenya's wildlife and pastoral activities, but not good for farming, were taken over by emigrants for crop cultivation and there has been additional loss of biodiversity [5].

In addition to preserving sustainable ecosystems, it is certain that existing forest land has to be preserved for both economic and social reasons. The loss of biodiversity is a threat to tourism and the employment opportunities it offers. Tourism accounts for one-third of Kenya's foreign exchange, a figure that has been increasing. With habitat degradation/loss and declining biodiversity as some displaced wildlife migrates to Tanzania, a share of tourism and the foreign exchange income it brings to Kenya could be lost to Tanzania and cause economic stress for the Kenyans. To prevent this, there have to be protected areas with ample buffer regions, where zoning legislation is enacted to relieve ecosystem stress and vigorously enforced if the laws are disregarded. Ideally, such laws would require that old stands of forests be protected and that a tree to be felled be cleared of vines binding to other trees so that they are not also uprooted. To further protect forest ecosystems, the laws should also require that felled trees be moved by an overhead cable to where they will be loaded on trucks or unloaded into waterways for transport to mills. An added requirement should be to plant saplings where a tree is felled. Finally, the laws should ban clear cutting in order to maintain a vegetation cover that supports natural seeding on a forest floor, and to prevent soil erosion.

8.3.1.3 Pollution

Chemical pollutants in air, water, soil, and potentially toxic heavy metal pollutants weathered from rocks diminish the health and viability of ecosystems and the communities they nourish. This is a serious problem in middle- and low-income developing and less developed countries but one that affects developed nations as well. The stress on citizens is biological as toxins can access the body and attack critical organs. The stress is mental as well from the suffering of family and friends that fall to pollution-induced illnesses. Governments with the legislative will and/or economic resources can force polluters to install and use the necessary technology to greatly reduce their contributions to ecosystem pollution loading. This is very important at urban centers with growing populations and high population density. Although the expense of cleanup of inherited sources of pollutants is high and the remediation process slow, reclamation is possible in many ecosystems. However, in the case of high-level radioactive pollutant sources such as at Chernobyl or Fukushima Daiichi, released radioactivity and polluted juxtaposed regions, it is too dangerous to clean up the radioactive pollutants from contaminated areas making it necessary to evacuate resident populations. Containment is the *modus operandi* at the radioactive facilities together with a quarantine of the lands that register levels of radioactivity dangerous to humans until the decay of radioactive elements reduces radiation to non-health threatening levels. This can take 100–1,000s of years or more. At Chernobyl and Fukushima Daiichi, stress from radiation sickness and from seeing the anguish of others is added to by the physical displacement from homes and land and the possibility of suffering illnesses years into the future from the initial high exposure to radiation.

You, the reader, have already learned about air pollution from coal combustion such as Hg and other heavy metals, Pb and other heavy metals from smelters, Pb from battery factories, and about the biological stress that the ingestion and bio-accumulation of one or a combination of heavy metals can put on the human body. These chemical elements may have short- or long-term chronic effects that impart stress on people, especially babies, young children, and pregnant women. Methane and other gases generated from surface waste disposal sites close to urban populations can be dangerous. Airborne fine-size particles (less than 2.5 μ) in the atmosphere are a risk to the human respiratory system as they access and accumulate in the lungs. Certainly smog, a product that forms from chemicals emitted from industrial processes mixed with emissions from cars and trucks in heavily populated cities can damage the bronchial system and harm cardiac health if inhaled over several days. As described in Chap. 7, smog often develops over cities with geographies that favor meteorological conditions for smog to linger for a period of days or even weeks. When near ground level airborne chemical threats such as these recur in a city, governments have to determine how best to protect its people from the stress brought on by smog-caused illnesses. This can be by requiring the use of efficient control technologies at emission sites, by legislating (for the near future) stricter emission control for internal combustion engines, by limiting the access by cars and trucks to areas under smog threat, and by keeping people at home for the duration of a smog condition.

In areas served by water collection, treatment, and distribution plants, the finished product is regularly analyzed chemically and biologically to maintain quality control and safeguard the health of users. When an anomaly is detected, scientists and water engineers review the water cleansing process step by step, determine the origin of the problem, and fix it. Many populations draw their water from municipal or individual wells. These should be tested regularly lest bacteria or potentially toxic metals are present that can cause stressful health threatening conditions such as diarrhea. If chemical or biological tests show the well water to be contaminated, scientists work on the problem until a source or more than one source is identified. Possible sources can be targeted from the land use history of a location. In essence, toxic pollutants in aquifer water can originate from the following: (1) untreated or incompletely treated industrial effluents that run off onto soils and rocks and seep into aquifers; (2) leakage into aquifers from surface or buried waste disposal sites; (3) leakage from surface or underground chemical storage tanks into aquifers; (4) runoff from commercial animal husbandry operations (cattle, poultry, sheep, goats), and from agricultural chemicals (pesticides, herbicides, fertilizers) that seep through soils and rocks into aquifers; (5) from chemical reactions between water and aquifer rocks; and (6) from acid mine drainage onto land that can access aquifers and surface water. Pollutant runoff that reaches waterways can disrupt their ecosystems. Once identified, scientists can work with engineers to determine how to solve the pollution source problem. If the pollution cannot be stopped at its source, the ideal solution is to move the contaminated water to collection and treatment facilities before releasing it for public use. However, this is not practical at some locations because of the economics of building, staffing, and maintaining a treatment facility and its infrastructure.

Lastly, a dangerous pollutant that one cannot see, hear, smell, touch, or taste is the radioactive gas radon (Rn). The gas can accumulate in lower level of homes if built over rocks that contain small concentration of uranium (Chap. 4). Inhalation of Rn over time is believed to be one cause of lung cancer. Here, mental stress comes on rapidly when citizens learn that their families have been inhaling a radioactive gas that can cause lung cancer and that there must be a renovation that eliminates the danger from a home to make it safe. Measurements of Rn emitting from soil/rock at possible home sites is now required by many municipalities before issuing permits for construction.

Clean air, safe and sufficient water, and fertile and untainted soils to grow enough and good quality food crops make up the triumvirate of requirements to sustain life for the world's growing population. Maintaining these earth phases pollution free keeps populations free of stress from health problems.

8.3.2 Natural Hazards—Primary, Triggered

8.3.2.1 Earthquakes

Earthquakes are unpredictable as to when they will strike, where they will strike, and with what magnitude and motion (shaking, jarring, rolling) they will have. Major earthquakes on land are relatively few annually and do not generally revisit the same location in the human timeframe. Nevertheless, people living in areas that have suffered from a destructive earthquake live with anxiety although there is a low probability of a major repeat event during their lifetimes. When there is a low magnitude ground motion and no damage, populations are not flustered but are aware that the Earth is not static. Stress sets in after a high-magnitude earthquake that causes structures to collapse, with people killed and injured. Initially, citizens panic even as disaster response teams arrive to find and save survivors and relief teams provide water, food, shelter, medical attention, and counseling for those that seek it because of post-traumatic stress from the loss of family, and loss of homes and livelihoods. The aftermath raises questions of why buildings and infrastructure failed and what improvements can be made to building codes to minimize death, injury, and damage in a future high-magnitude earthquake. The stress on populations is long lasting during cleanup operations and reconstruction. The stress will lessen as normality returns but this may take years to reach as exemplified in Haiti where a killer earthquake hit the Port-au-Prince area in January, 2010 but where cleanup and reconstruction continues in 2014 as populations live with chronic anxiety and a cholera epidemic.

8.3.2.2 Tsunamis

Tsunamis can be the most damaging of earthquake-triggered events. As described in Chap. 4, they are generated from earthquakes that occur beneath the ocean floor. There are tsunami warning systems in many oceanic sites that are activated as an ocean earthquake hits when a fault ruptures and causes a vertical displacement of the ocean floor as one side of the fault moves up or down many meters with respect to the other side. The earthquake motion and results that may follow causes great stress for populations and the possibility of an incoming tsunami intensifies their high anxiety. In place, warning equipment predicts where, when, and what height wave can be expected to impact shorelines. Depending on this prediction, a population stress level can increase further although the warning gives citizens time to flee from the coast inland and to higher ground. Where these warning systems are not in place, people can be killed and injured, coastal towns and installations destroyed, and ecosystems ruptured with torn out vegetation, animals drowned, and saltwater-covering agricultural fields. This intensifies the post-traumatic stress suffered by the earthquake impacted population. This was the case in December, 2004, in the southwest Pacific Ocean where a tsunami hit Sumatra, Indonesia, Thailand, India, and Sri Lanka and killed 230,000 people (about 200,000 on Sumatra). This region now has a tsunami warning system.

More recently, in March, 2011, a major earthquake and a mega-tsunami it generated hit northeast Japan, destroying 164,000 structures and killing, injuring, and displacing thousands of people. The destruction it wreaked on population centers was compounded severely by the damage caused when the tsunami crashed into and disabled a nuclear power facility destroying its coolant system (Chap. 4). Without coolant, the reactors overheated and exploded ripping through containment roofs and allowing the release of dangerous radioactive isotopes (e.g., iodine I-131, cesium Cs-137, and Cs-134 as well as strontium Sr-90, and three isotopes of plutonium (Pu-239, 240, 241) and particles into the atmosphere that fell out on land and laterally discharging radioactive coolant into marine waters. Habitation is prohibited and crop land cannot be worked within a 20-km radius of the power plant because of the loading of radioactive isotopes into the area. The Pacific Ocean is damping the radiation by dilution and captured fish are monitored for radiation before being released for sale. More than 250,000 displaced people now live in temporary housing. As a result of these events, citizens that survived suffer from stress-related physiological and mental health problems (post-traumatic stress disorders) knowing that there are no towns and villages to return to and that it might be many more years before they can move into permanent housing. Others suffer stress because of economic setbacks such as losing livelihoods or export markets. For example, Korea halted the import of food fish from five prefectures in Japan in areas associated with the Fukushima Daiichi disaster although radiation levels in the fish are below internationally acceptable concentrations. Anxiety also gripped much of Japan to the south as air currents transported a diluted radioisotope load in their direction. The shutting down of all Japanese nuclear power facilities followed so that the electrical grid could not supply the necessary power that was needed for all domestic, manufacturing, and industrial uses. This led to a slowing or shutdown of many operations, stressing the country's economy. In 2013, a few nuclear-powered electricity generating plants were allowed to come back onto line. The future for nuclear power stations in Japan is undecided given the gravity of the Fukushima Daiichi disaster.

8.3.2.3 Volcanoes

Volcanoes give signs that there is the possibility of an imminent eruption but there is no indication when exactly an eruption could happen, no indication on the magnitude of an eruption, and no indication of where on the volcano the eruptive burst may occur (at its summit, from its slopes). From the history of a volcano, the type of eruption can be predicted such as a quiet lava flow or as a violently explosive vertical or lateral eruption of molten magma ejecta, glowing ash and incandescent gases, and rock fragments, or a combination of these. Geophysicists, volcanologists, and geologists observe a volcano, assess measurements they take with sensitive equipment, review the topography, and evaluate a volcano's history before they decide whether to make a recommendation for evacuation away from the reach of a potential volcanic eruption. With greater signs of possible eruptive activity, citizens' stress levels heighten as they await reports from scientists of the need to evacuate or not. In lightly populated areas on and around a volcano with a modern history of recent activity, people are aware of but not overly concerned by the possibility of evacuation to escape a potential eruption. Yet, they suffer greatly if an eruption takes place and devastates their villages and work places. However, where there is a history of a primary event such as a nuee ardente (a glowing, fast-flowing mix of ash flows with hot gases) that has wiped out entire cities, or triggered events such as mud flows or lahars (hot mud flows) that have buried cities, the populace is alert to the possibility of this volcanic activity and prepared to evacuate. In general, people are not overly anxious about the proximity of a volcano until it shows activity.

8.3.2.4 Floods

In contrast to earthquakes and volcanoes, floods are predictable. With a flood warning system is in place, scientists can determine when flood waters will arrive at a given waterway location and the height to which water is likely to rise (see Chap. 4). Because populations know what is coming, they are not stressed to the panic level that can by a sudden event such as a flash flood when a deluge of rain falls over a short or sustained period. With a flood alert, people have time, perhaps days, to prepare to sandbag homes and increase the height of levees to hold back flood waters or to gather important papers, treasured family items, valuables, and move items to upper floors if possible before responding to a call for evacuation. Certainly, the loss of property and in agricultural zones, crops, to flood waters stresses the affected people and economies of impacted areas. Unfortunately, flood prediction, warning systems, and flood control structures are lacking in some regions in developing and less developed countries so that threats of seasonal storms with heavy and sustained rains stresses citizens during these times in many densely populated areas such as in China and the Indian subcontinent.

8.3.2.5 Weather

Anxiety assaults society when there is a warning of extreme weather events such as tropical storms that can evolve into hurricanes (typhoons, monsoons), tornados, drought, heat waves, major snow storms, and frigid weather conditions. Stress in the citizenry continues with the onslaught of an extreme weather event and may increase depending on what the event has wrought upon a population. However, weather predictions have degrees of uncertainty. These include when an event will happen, the path it will follow, the areas it will impact, its destruction potential from wind energy/speeds/motions and coastal storm surges, the amount of precipitation it may bring, temperatures expected, and how long an event and its aftermaths will last. The uncertainty of weather predictions and changes is stressful for people following reports when potentially destructive storms are forecast. People fear storms that can kill and injure people, cut utilities or other services, and make accessibility to food, water, medical care, and search and rescue efforts difficult to impossible, at least for the short term.

8.4 Human Actions/Inactions that Intensify and Concentrate Impacts of Natural Hazards

The effects of natural hazards and stress they cause in populations are often strengthened by direct or indirect enhancements from human activities. Prime examples of this have to do the emission of CO_2 to the atmosphere through the combustion of fossil fuels, the global warming it support, the cutting of forests (e.g., in the Amazon) for wood and land use, and water. Many countries are experiencing increased frequency, intensity, and reach of floods that put more people and property at risk. This is thought to be caused by global warming and the more moisture it puts into the atmosphere that is moved landward and released as heavy, sustained rains by storms. In other scenarios, inaction or bad decisions in terms of managing development in the upper reaches of a drainage basin can cause or exacerbate flood conditions for populations downstream. Extreme weather events that are affecting more populations worldwide are the results, at least in part, of human activity fueled climate change. The effects of such events are magnified by human populations concentrating in ever increasing numbers in coastal zone urban centers or near major rivers. This increases their risk of harm from seasonal high-energy tropical storms (and occasional earthquakes or triggered tsunami). Certainly, diversion of rivers by governments to stimulate economic growth or provide safe water to parched populations can lead to social and economic stress for those populations that lose their source of water and disrupt ecosystems or essentially bring them to extinction. Inaction by not keeping building codes up to date or enforcing them if they are current can be especially damaging where there is high density of citizens living in earthquake zones. Structures have collapsed worldwide away from earthquake zones because of dated building codes or unenforced codes and corruption-permitted shoddy construction. Similarly, inaction by not establishing zoning laws that prevent construction in dangerous sites such as flood plains and landslide prone hillsides puts people and property at increased risk when a recurring event impacts such an areas. In all cases, loss of life, injury, and loss of property cause extended periods of stress for the populations that struggle to recover from a natural disaster.

8.5 Stress Reduction

Stress reduction of biological and physical elements, economic and sociological conditions, and political factors listed in Table 8.1 can be achieved. Simply reverse the causes that bring on stress, steps more easily proposed than done and unlikely to be implemented worldwide. Most nations will agree in theory to work individually and together to reduce factors that stress their populations. However, when economic, environmental, social, and political actions unfavorable to governments are called for to be instituted, governments will stall and put up barriers to progressive reduction of many of the sources of stress. In the instance when an environmental treaty (the Montreal Protocol) was agreed to by 191 nations to save the ozone layer and thus reduce the health problems UV rays caused, there was a slowdown and halt of production of chlorofluorocarbons and stoppage of their release into the atmosphere. This arrested the degradation of the Earth's protective ozone layer and originated the start of its reconstitution. There was little if any economic impact on signatory nations, a fact that led to the success of the treaty. Conversely, the international efforts at several meetings since 1992 to draft a binding treaty to slow global warming and climate change by each country reducing the mass of CO_2 and other greenhouse gases its industries emit to the atmosphere have not been successful. The reason a treaty has not been agreed to and signed is because of economic restraints and need by developing countries with growing populations to maintain and strive to increase industrial and manufacturing employment levels. The CO_2 emissions in the USA and other industrialized nations have been reduced dramatically during the past few decades. Brasil as a developing country cut its 2010 CO_2 emissions back to 1990 levels. This went hand in hand with a dramatic reduction in human encroachment into the Amazonian jungles, thus retaining more of an important CO_2 sink. However, these reductions have been negated by growing increases in CO_2 emissions mainly from China (the global leader) but also from other developing countries as they rush to industrialization and economic development.

8.6 Afterword

Stress in our growing populations can be relieved to a certain degree by adaptation to changing conditions as they evolve. An adaptation may be as "simple" as installing a dam for flood control or as complex as moving a coastal village, town, or city inland as protection against recurring, and stronger storm surges or building an extensive and expensive protective sea wall to hold back the main force of a surge. Adaptation to changing climate is a priority in agricultural zones to assure food security, and adaptation to changes in water availability is essential. Adaptation of international, regional, national, and local populations to change is discussed in the next (and final) chapter.

References

1. Siegel FR (2010) The exploding population bomb—societies under stress. Washington, DC 225 p (Self-published)
2. webMD (2012) The effects of stress on your body. www.webmd.com/balance/stress-management/stress-management-effects-of-stress. Accessed 16 Mar 2014
3. United States National Institute of Health (2014) Fact sheet on stress. www.nimh.nih.gov/health/publications/stress. Accessed 16 Mar 2014
4. Schnederman N, Ironson G, Siegel SD (2005) Stress and health: psychological, behavioral, and biological determinants. Annu Rev Clin Psychol 1:607–628
5. Oteno J (2010) Vegetation loss threatens to push wildlife species into extinction. Daily Nation, Kenya, 1 Nov

Chapter 9
Progressive Adaptation: The Key to Sustaining a Growing Global Population

9.1 Adaptation

Adaptation is an evolving long-term process during which a population of life forms adjusts to changes in its habitat and surrounding environments. Adaptation by the global community as a unit is vital to cope with the effects of increasing populations, global warming/climate change, the chemical, biological, and physical impacts on life-sustaining ecosystems, and competition for life sustaining and economically important natural resources. The latter include water, food, energy, metal ores, industrial minerals, and wood. Within this framework, it is necessary to adapt as well to changes in local and regional physical conditions brought on by natural and anthropogenic hazards, by health threats of epidemic or pandemic reach, by social conditions such as conflicts driven by religious and ethnic fanaticism, and by tribalism and clan ties.

9.2 Adaptation to Population Changes

Although the rate of global population growth is declining and is expected to fall to the replacement level by mid-twenty-first century, it is still increasing by about 75–80 million people annually (see Chap. 1). A grand part of the growth is taking place in Africa, Asia, Latin America, and the Middle East that together already accounts for a 2014 population of 5.9 billion of the 7.2 billion people worldwide. Conversely, populations are contracting or stable in most of Western and Eastern Europe, and Japan. Together with the United States that has a stable population, this latter group today is home to the other 1.3 billion people. Thus, although there are two population situations to adapt to, both have common problems of sustainability to address.

© The Author(s) 2015

F.R. Siegel, *Countering 21st Century Social-Environmental Threats to Growing Global Populations*, SpringerBriefs in Environmental Science, DOI 10.1007/978-3-319-09686-5_9

   The principal problems with growing populations do not involve space although population density is a problem unto itself for reasons discussed in previous chapters. The main problems are how to nourish people with food and water. The chronic malnutrition that about 1 billion people suffered from in 2013 is likely to grow in number in some regions due to global warming/climate change because humans cannot adapt to less food if they are already at subsistence rations. For example, the 2012 population in sub-Saharan Africa is 902 million people. The population is projected to increase to about 1.25 billion in 2030, an increase of about 38 %. Within the same time frame, the United Nations estimates that acreage under maize cultivation in the region will decline by 40 % because of heat and drought brought on by climate change. The loss of arable land for food production can be countered in sub-Saharan Africa if marker assisted hybridization of maize or maize genetically modified to withstand heat and drought come onto the seed market together with modified seeds for other food staples and if African nations that do not now accept GMO seeds do so in the future. If not, nations favored for food production by climate change will have a moral obligation to provide food staples to people in nations with declining food production at accessible costs based on their economies. It is clear that what happens in sub-Saharan Africa and other regions with declining cultivation acreage or that will bear other effects of climate change (e.g., drought, shifting rain patterns) will affect the rest of the worldwide community politically, economically, and socially. The earth's problems that associate with global warming/climate change will be further discussed in the last section of this chapter.

   Water is the staff of life. It keeps the body hydrated and is necessary to grow food crops, hydrate food animals, and grow feed grains. Chemically or biologically polluted water does not serve these ends. If ingested, contaminated water can result in sickness as discussed in Chap. 2. Water stokes industry and manufacturing as well, thus keeping economies in many countries contributing to a population's well-being by providing employment, goods, and services. Ideally, these businesses contribute their fair share to a tax base that supports social needs (e.g., education, healthcare, maintenance of infrastructure). Factory owners adapt and plan against water shortages by having a water recycling system in place but may also slow or stop production until operational water conditions return. Citizens with a reliable water supply can adapt to periods of water shortage by limiting use according to mandates by government officials but still have water for basic daily needs. However, persons in nations with a chronic per capita water shortage may not have this option to serve their daily needs unless water is imported or new water sources are found (see Chap. 2). If imported water is not an option to meet immediate essential needs, an alternative adaptation for people (and animals) is to try to reach a location where water would be available to them. With growing populations, per capita water availability is greatly diminished (Table 2.2), water shortages become endemic, and people are at risk of existing at subsistence levels or dying. Most at risk from the lack of a basic water ration are pregnant women, infants, young children, and old people. Water wars are a future possibility as nations battle for their peoples' survival unless political differences are set aside and projects are supported to develop and share water sources. In a welcome effort, Jordan, Israel, and the Palestinian Authority signed a memorandum of understanding in the World Bank, December 2013, with specific aims: (1) produce millions of cubic meters of drinking water for a water-deficient region; (2) pipe 200 million cubic meters of water annually ~ 180 km (110 mi) from the Red Sea to the Dead Sea; (3) build a desalination plant at Aqaba that would supply water to Aqaba and Eilat; (4) the Israeli Water Utility would supply 20–30 million cubic meters of drinking water to the Palestinian Authority for the West Bank population at a reduced cost; and (5) there would be an inflow of water to slow and in the future perhaps abate and reverse the shrinking of the Dead Sea. Funding for the estimated \$400 million, 5-year project will come from the World Bank, donor nations, and philanthropic groups.

As the global population increases and more people in developing and less developed nations have more disposable income, there will be a growing draw on natural resources other than water and food to service their industrial, agricultural, and manufacturing needs and wants. Competition can force economic wars among national and multinational corporations for the resources necessary to provide goods and services and thus drive up prices for resources. Industries and manufacturing units that cannot compete economically for natural resources will shut down, thus contributing to unemployment and downturns in economies because of falling domestic demand. To keep order in the increasingly interdependent world economy, accommodation for shared natural resources (or substitutes for them) at affordable prices is the adaptationnecessary. This can be mandated by the World Trade Organization backed by other practical-minded international groups.

Another adverse effect of growing populations that is a national resource that can be lost at the expense of some countries to the benefit of others is brain power. This brain power has been cultivated at excellent universities in developing countries, often times at little or no cost to students (e.g., in medicine, science, engineering, economics, the arts) who attend and graduate in increasing numbers. Where there are too many well-educated professionals but lack of employment opportunities for them in their fields of expertise, educated people have the option of relocating to another country that can nurture and use the expertise. Many adapt to the employment problem by taking up this option. This may mean moving from a developing country to a developed country or from a less developed country to a developing or developed country. Ultimately, this loss of citizens with special skills can hurt a country. To counter this brain drain or reverse it, a country can adapt by investing in its future to create programs and conditions that keep talented professionals home, or if they have emigrated, entice them to return. China and India are examples of countries that have successfully taken this tact.

When there are increases in a population because of immigration, problems can ensue between immigrants and a general population. Adaptation to diversity and the multicultural experiences it brings to a community is often not a comfortable change. The antipathy of some in a host country is based on slowness of the immigrants to learn the language and inability of host country citizens to understand what immigrants are saying among themselves. This makes citizens feel uneasy.

Some view immigrants as a threat to their own or a family member's employment or advancement. Race difference is a factor that some cannot readily adapt to as is ethnicity with its traditions and customs unfamiliar to the general public. Religion can be divisive if adherents to its beliefs engage in acts of hatred detrimental to the host country fueled by fundamentalists and zealots who interpret religious writings as giving them license to commit crimes or absolving them of the crimes. Sadly, many citizens paint an entire religious community with the taint of the relatively few evildoers. Adaptation to diversity is essential for our earth's citizenry with joint efforts by all to resolve worldwide issues (e.g., global warming/climate change) so as to become the keys to providing a sound future for coming generations. There has to be a shared attack on global threats, no matter what the language, race, ethnicity, or religious beliefs are, no matter social or economic status, no matter whether a threat affects less developed, developing, or developed countries.

9.3 Adaptation to Natural Hazards

Adaptation is a progressive process when dealing with natural hazards because as each type of natural hazard impacts global communities over time, lessons are learned from each one that give direction to the methods of adjustment. Adaptation to living where hazards can be expected to strike and where populations continue to increase is dependent on what we learn from the study of past hazards. We can use this evaluation of measured and observed data to minimize the immediate effects and aftermaths of hazards and protect citizens from injury, death, and from damage or destruction of property or infrastructure when hazards strike in the future.

9.3.1 Earthquakes

In areas prone to earthquakes, we know that earthquakes do not kill and injure people but that collapsing buildings and infrastructure do. Earthquakes are not predictable so that there is no adaptation by a timely evacuation to minimize deaths and injury. However, building structures to make them more earthquake resistant can save lives, reduce injuries, and protect property. Thus, after a high-magnitude earthquake, forensic engineering teams come to assess the damage and determine where and why damage and destruction took place within the context of the magnitude of an earthquake, the type of motion it originated (shaking, jarring, rolling), its duration, the area it affected, and the geologic properties of rocks underlying structures' foundations. Hazard assessment teams also evaluate other factors that contributed to additional damage such as ruptured gas lines that feed fires and ruptured water lines that inhibit fire control. The engineers establish how construction can be improved in the future in terms of construction techniques and materials to prevent the types of collapses and utility failures they investigated.

Municipalities revise building codes accordingly to direct reconstruction and future building projects. Where possible, structures that withstood an earthquake with minor or no visible damage should be retrofitted to improve their resistance to the next "big one." With each event, we gain more data on how to better construct earthquake-resistant structures and alter building codes to more stringent specifications. In theory, this adaptation to an irregularly recurring global event is good, but in practice it is most applicable to nations with the economic resources for reconstruction according to revised building codes and where there is no corruption to allow a bypass of the code. The same can be stated for retrofitting to give more resistance to earthquakes to existing structures. Many developed nations and nations rich in commodity exports (e.g., oil) have a moral obligation to donate funds, material, and expertise to help citizens in economically disadvantaged nations recover from a destructive earthquake. Some commodity-rich and economically sound nations do not do so directly, whereas others, big and small, rally to help disaster victims. For example, immediately after megatyphoon Haiyan devastated many regions in the central Philippines in 2013, Israel sent 250 medical doctors and nurses and field hospitals to help Philippine citizens recover from the impacts of the typhoon.

9.3.2 Volcanoes

As discussed in an earlier chapter, volcanoes are predictable in terms of becoming active by emitting wisps of smoke, bulging on a slope, warming of the soil or nearby pond or lake waters, emitting increasing concentrations of gases, and showing increased low-frequency seismicity. However, this activity does not always result in an eruption. A marked increase in measurements and observations, especially the low-frequency seismic activity, suggests that an eruption is imminent. Adaptation to living and working on or near a volcano means investing in equipment to monitor volcanic activity and listening to alerts from scientists monitoring its activity and being ready to evacuate by gathering important papers and precious mementos and prepared to load into transportation for evacuation to safe locations. Governments adapt by charging geologists to map out areas considered as high-, moderate-, and low-hazard zones in the volcano environs. Geologists do this by studying rocks deposited from past eruptions and assessments of the topography. Municipalities then pass zoning regulations applicable to the hazard level.

9.3.3 Floods

Governments have adapted to repeated periodic flooding in areas by creating flood control systems described in Chap. 4. Dams hold water during times of heavy and/ or extended rainfall and release any overflow into channels that move water away from urban or rural population centers. Levees increase the volume of water that can move through a channel, thereby keeping it from spreading into populated areas and cultivated farmland. For smaller waterways that flow through cities, municipalities may invest in deepening, widening, and straightening channels as well as erecting walls so that more water can flow through the area more rapidly without coming out of a channel. Governments define zones on flood plains according to a recurrence interval of damaging floods (e.g., 100 years) as being off limits for residential and factory/plant construction. As much as we plan to adjust to living in an area prone to flooding, there is always the possibility of a megaevent that can overcome in situ control systems. Therefore, as described Chap. 4, governments adapt to this possibility by installing flood prediction equipment in drainage basins to provide warning to those at risk from rising and sometimes raging waters. The warning gives people time to gather important documents and personal treasures and evacuate to safe areas. The apparent increase in the frequency and magnitude of storms and resulting flooding in recent years is thought by many weather scientists to be related to global warming and the increased amount of moisture in the atmosphere from warmer oceans that gathers in clouds and precipitates during storms. This will be discussed further in this chapter.

9.3.4 Extreme Weather Conditions

Adaptation to extreme weather events such as an extended period of drought, heat waves, and frigid weather means preparation to wait them out. Some municipalities adapt to repeated, sometimes seasonal, times of short-term drought by storing a 3–6 month water supply in surface or underground reservoirs during periods of normal precipitation that can be tapped (conservatively) as needed. Others may plan to move water via pipes or water tankers from where it is plentiful to where drought conditions exist. Otherwise, to survive, people move as best they can to where they have access to water. In instances of years long drought, crops and livestock and other life forms may be lost. Heat waves can kill. Adaptation to heat wave conditions means that water has to be available to people to avoid dehydration. Where possible, homes should have air-conditioning or fans to keep people comfortable and municipalities should have cooling centers to which people can go. Personnel should check on senior citizens and escort them to cooling centers if necessary. Clearly, economically advantaged nations have the resources to give support to citizens during natural hazards such as these. These nations, international organizations, and NGOs have a moral obligation to help economically disadvantaged nations as is possible when hazard conditions such as these threaten populations.

The most extreme of weather conditions that can injure and kill people and destroy housing and infrastructure are tropical storms that evolve into violent hurricanes (typhoons, monsoons) by increasing wind speeds and sucking up moisture (water) as they track across oceans toward land. When these storms make landfall, they drive storm surges that can wreak havoc onshore communities, and as they move inland precipitate heavy rains that cause life-threatening and destructive flooding. These violent storms are destructive to coastal populations and island nations and have regional reach inland as they move along paths until they finally spend their energy or move out to sea. On November 8, 2013, the typhoon named Haiyan, the strongest recorded typhoon ever to make landfall smashed into the central Philippines killing more than 2,600 people, injuring about 12,500, and displacing almost 600,000 people. There was a 4-m ($\sim$ 13 ft) storm surge driven by winds measured at over 312 km/h (195 mi/h) with gusts reaching 380 km/h (235 mi/h). The typhoon flattened the city of Tacloban that was home to 200,000 residents, and there was major flooding inland. The weather alerts led to a government call for evacuation away from the predicted path of the storm, and about 1 million people followed the evacuation warning, surely saving many lives. Access to aid typhoon-ravaged areas was difficult, and there were shortages of water, food, and medical care for many evacuees for several days. The Philippine central government and local officials were not prepared to deal with a storm of this magnitude but help started arriving from many nations worldwide. There was a post-event concern of attending to sanitation needs of survivors to prevent outbreaks of diseases such as cholera, typhoid fever, hepatitis, and dysentery. If the Philippine government had adapted by adopting better policies with respect to response to high-category typhoons in addition to the call for evacuation, the impact of Haiyan would have been ameliorated. One would hope that this deficiency would be dealt with to limit the effects of future like disasters.

Evacuation to prevent injury and death in coastal zones that could be struck by high winds, heavy sustained rains, and storm surges is dependent on weather bureau forecasts and warnings from police, firefighters, or other government-authorized personnel. Homeowners adapt to hurricanes by securing roofing with additional nails or special fasteners as a retrofit precaution and by boarding up windows on structures before an incoming storm hits. Governments have adapted to the onslaught of violent high-energy storms by constructing seawalls of varying designs and heights to protect population centers by damping the force of storm surges. In China, for example, a seawall 6.72 m ($\sim$ 22 ft) in height and that has been heightened in the past protects Shanghai from the full damaging effects of high-category typhoons. As a result of rising sea level, the Shanghai seawall and other that protect coastal cities from being flooded by surges from high-energy tropical storms will have to be heightened to afford a greater degree of protection to people and property.

9.3.5 Wildfires

Wildfires can be a natural hazard when ignited by a lightening strike. However, most wildfires are started by human carelessness such as tossing a lit cigarette on a forest floor or failing to completely extinguish a campfire, or by arsonists. One may adapt to living in an area with a history of wildfires in two ways, neither of which is practical or promises 100 % protection. First would be to clear an area of vegetation in a 30-m (100 ft) swath around a dwelling or site for building. Second would be to build with nonflammable materials so that embers propelled during a wildfire could not ignite a structure. Adaptation to the advance of a wildfire would be to heed warnings to evacuate carrying a prepared case with important documents and other items of personal value. To delay evacuation by going back to retrieve something from then home can be fatal as it was for two people in a recent (June, 2013) wildfire that destroyed almost 500 homes in Colorado Springs, Colorado, USA.

9.3.6 Preparedness Against Natural Hazards

When there is a hazard event coming that calls for evacuation, responsible and often economically advantaged governments have adapted to the threat by designating evacuation routes, by providing transportation for people who need it, by having evacuation centers stocked with water and food, cots and blankets, basic medical supplies and medical personnel, and by having phone service available for people that need it. In the case of a primary or triggered hazard that happens with little or no warning (e.g., an earthquake, a tsunami, a volcanic mud flow), search and rescue teams should be ready to move in soon after dangerous conditions ease and they can move with safety. There should be medical attention to treat injured survivors, and stations set up as soon as possible to provide water, food, and other essentials available to those that survived with little or no physical hurt. These first steps at adaptation are the keys to survival. Recovery after a shock phase can be long and drawn out, depending in grand part on a nation's social and economic resources and physical and economic assistance from other nations, international institutions, and NGOs.

9.4 Adaptation to the Effects of Global Warming/Climate Change on Our Earth's Inhabitants

9.4.1 Cause of Global Warming

Global warming is a fact attested to by an overwhelming majority of the scientific community and unwaveringly supported by a February 2014 joint publication of the US National Academy of Sciences and The Royal Academy in the UK on the causes and evidence for global warming [1]. As noted in earlier chapters, during the past century, measurements show that the earth has warmed by ~ 0.8 °C (~ 1.44 °F). Global warming is an ongoing process that is attributed in grand part to a slow but continuous and increasing buildup of greenhouse gases in the atmosphere. The greenhouse gas most associated with global warming is carbon dioxide

(CO_2). A plot of the increase of CO_2 content in the atmosphere with time against the increase in global temperature shows an excellent correlation of one with the other. Additional lesser contributors include methane (CH_4), nitrous oxide (NO_2), and chlorofluorocarbons (CFCs). With the beginning of the industrial revolution and the increased use of coal as the principal energy source, the content of CO_2 in the atmosphere was 280 parts per million (0.028 %). The combustion of coal and later oil (petroleum) and natural gas emits CO_2 to the atmosphere. Initially, and for many years thereafter, the added greenhouse gases were taken up by vegetation for photosynthesis and was also absorbed by the oceans and other water bodies. This kept the atmosphere CO_2 close to the 280 ppm pre-industrial level. However, with increased industrialization, the need for electrical power, and the use of internal combustion engines, the amount of CO_2 generated was greater than what could be absorbed by nature and the content of CO_2 in the atmosphere increased. During June 2013, its concentration reached more than 398 ppm, an increase of over 40 % over the pre-industrial concentration (Scripps Institute of Oceanography Mauna Loa measurement). The increasing CO_2 content, other greenhouse gases, aerosols, and particles acted as a media that admitted sunlight (heat energy) to the earth's surface but did not let all of the heat escape back into the atmosphere. This abets global warming. In the past two to three decades, the rush to industrialization in developing countries (e.g., China, India, and Brazil) and their growing power needs and vehicular use has thwarted the implementation of international agreements to reduce emissions from coal-fired power plants, other industrial and manufacturing operations, and the transportation sector.

9.4.2 Effects of Global Warming

9.4.2.1 Rising Sea Level: Adaptation/Mitigation

A direct consequence of global warming is sea level rise (SLR) caused by the progressive melting of icecaps and ice sheets in Greenland, the Arctic, and Antarctica, and of mountain glaciers in the Himalayas, the Alps, the Rocky Mountains, and the Andes. The $\sim$20-cm ($\sim$8 in) sea level rise during the past century may see a rise of another $\sim$50 cm ($\sim$20 in)–1 m ($\sim$39 in) during this twenty-first century. One-third of the rise would be from the expansion of warmer sea water, one-third from icecap and ice sheet melt, and one-third from mountain glacier melt [2]. In 2012, other researchers used computer models on existing data and proposed that 50 % of sea level rise between 1903 and 2007 was from glacial melt [3]. Following the same line of investigation, other scientists studied satellite data and ground measurements from Alaska, the Canadian Arctic, Greenland, the southern Andes, the Himalayas, and other high mountains of Asia and estimated that glacier contributions to sea level rise from 2003 to 2009 was 29 % and together with ice sheet melt explained 60 % of SLR [4]. A publication in 2012 estimated that ocean thermal expansion 0–300 m deep and 300–700 m deep contributed up to 35 % to sea level rise [5]. These latter two estimations are in line with the IPCC prediction for melting ice and ocean thermal expansion contribution to the estimated rise of sea level by the end of the century [2]. With a rise in sea level, marine waters encroach on land. As the rise continues, possibly at an increasing rate, it threatens habitation in lowlying islands, coastal villages and farmland in lowlying zones, and heavily populated cities worldwide settled on inshore terrain close to sea level (e.g., Bangkok, Ho Chi Minh City, Jakarta, Manila, Miami, New York, Boston, Buenos Aires, London, Rotterdam).

Rising sea level and warming of ocean waters have other ramifications that affect coastal communities as well as inland areas. As explained in Chap. 7, the warmer surface water releases more water vapor with heat energy into the atmosphere. When the water vapor molecules condense in clouds, heat energy is released. This energy gives more force to tropical storms as they form, track to shore, and move inland, or storms that move close to and along a coast. These storms may transition to hurricanes (typhoons, monsoons) with the violent winds that cause destruction, and heavy rainfall that triggers flooding if they move onto land. We recognize that rising sea level means that tropical storms that impact a coast with storm surges have a farther reach inland with their destructive energy that is more pronounced when the surge occurs at high tide. The surges also saturate farmland they reach with salt water that harms crops. They also carry salt water into fresh water marshes and ponds, thus disrupting ecosystems there. The increase in the number of these extreme weather events and the increase in violence and destruction they wreak on land compared with like weather events in the recent past (e.g., during the past 30 years) strongly suggest that they are fueled to a significant degree by global warming.

There are two possibilities for adapting to the effects of rising sea level on coastal urban centers, one impractical, the other very costly but doable. The impractical adaptation possibility is to move at-risk population centers inland, out of the reach of the destructive tropical storms. This does not lessen the threat of flooding. The move is possible in some cases where land is available, but such a move is not economically feasible. One practical but costly adaptation to mitigate encroachment from sea level rise and the effects of tropical storm surges is to surround cities at risk within place seawalls 2–3 m higher than recorded high tides or higher depending on historical records and contemporary published data. The walls can have a concave configuration so that surging waves lose energy when their lower parts hit and are curled back on themselves damping some wave energy or there can be a different configuration best for the site(s) to be protected. Similarly, gates buried at strategic locations where there is ship access to consider can be built to be hydraulically driven so that they can rise from a near shore seabed site to mitigate the effects of storm surges. Both techniques have been used at different global locations.

9.4.2.2 Food Security: Adaptation to Climate Change by Land-Based Agriculture

We have read that climate change affects land-based agricultural production, both for crops and animal husbandry. The warming climate at higher mid-hemispheric latitudes and at higher altitudes does not favor the growth and normal yield and/or quality of many crops. Depending upon the degree of climate change and the linked change(s) that may follow it, farmers can adapt in several ways to maintain or increase crop yield and nutrition value. For example, when warming starts diminishing the productivity of a traditional crop, farmers can sow crops that are known to grow well in warmer temperature and give a satisfactory economic benefit. However, new groups of weeds, pests, and diseases will migrate to the warmer growth environment and will have to be dealt with in order to protect the new crops.

Where the effect of global warming reduces water supply for rain-fed agriculture, for crops irrigated with surface waters, and for groundwater-irrigated crops when aquifer recharge does not balances discharge, agriculturalists can adapt in two ways. First is the use of a more efficient irrigation method that delivers water directly to a growing plant (e.g., drip or focused irrigation). This minimizes runoff and loss to evaporation. Second and similar to what was mentioned earlier is to sow a crop that needs less water to thrive and that delivers a good yield, good-quality product.

Another result of global warming for some farmlands is a longer growing season. In this situation, growers can adapt by planting earlier and have the possibility of double cropping. They can also grow a cultivar that is later maturing and that gives a product that brings a good market price. However, switching to new crops in a warmer growth environment means that there will be an invasion of a new set of weeds, pests, and diseases to ward off.

In any efficient operation, and as emphasized in earlier chapters, farmers adjust to a changing growth environment for a given cultivar by applying the optimum amounts of fertilizer and other agricultural chemicals as might be needed that nurture and protect it most effectively. This reduces agricultural costs and lessens runoff of these chemicals to ecosystems where they can be harmful.

Global warming can bring on abnormal weather extremes that affect agricultural productivity. In these cases, farmers have to plan ahead based on recent history of these conditions in their regions. Drought, heat waves, and long-term rain or heavy rain in a short time present problems for both cultivars and food animals. Periods of less than average precipitation may last months or years. Depending on the amount of the deficit precipitation, adaptation can include storing water in reservoirs and cisterns during times of rainfall to be tapped during a drought to sustain food animals and crops during a short-term, not too severe drought. There is also the option of trucking in water to sustain livestock. Long-term droughts when precipitation deficits are high take their toll on plants and animals to the detriment of agriculture in a region especially when accompanied by heat waves. They have caused recent disasters for crops and food animals on all continents less Antarctica.

Farmers either wait out the "bad times," change the type of cropping they do, the livestock they tend to, or change careers.

The adaptation from crops that have been grown successfully before the effects of global warming reduced yields and quality of a harvest, to those "same" crops that can grow successfully under the advancing warming changes just described generally means that hybridized species have to be developed and used as warming increases at a location and slowly tracks to higher latitudes and higher altitudes. Thus, growers turn to plants that are created by hybridization as described in Chap. 3: traditional methods and marker-assisted selection methods within the same species, and genetically engineered (-modified, -manipulated) methods using different species. Hybridization is a slow process, sped up markedly by genetic engineering, a method that yields foodstuff not accepted by the European Union and many nations outside the Union, especially in Africa. Bred species are developed to carry one or more characteristics that favor crop resilience against the effects of climate change. These include resistance to disease, weeds, and pests, and tolerant of drought (water stress), heat, short-term inundation, and short-term saline exposure (see Chap. 3). Hybrids have also been developed to give higher yields and more nutritious crops. Thus far, research has been focused mainly on improving seed for world staples such as rice, maize (corn), wheat, sorghum, and soybean.

There have been great successes where hybrid crops were agriculturalists' adaptation so that the possibility exists that we can feed the earth's growing populations and reduce chronic malnutrition. When this is coupled with the opening of additional arable acreage and the use of improved farming methods for seeding, watering, and harvesting, global food security can be strengthened for the existing world population and the future generations on earth. However, this will require economic and technical input by developed nations and international groups. Without basic sustenance, people will have less resistance to diseases and there may be local or regional population crashes if diseases evolve into epidemics or pandemics that invade susceptible populations.

9.4.2.3 Adaptation of Marine Fisheries and Aquaculture to Climate Change

Warming of the open ocean water, enclosed aquaculture operations in ocean waters and on land water bodies has affected marine fisheries and marine and estuarine aquaculture that grow food fish and shellfish, and lakes that sustain fisheries. In marine fisheries worldwide (e.g., in the north Atlantic, off the coast of Peru, off the coast of the Philippines), some food fish or fish captured for other purposes (e.g., to use in pet food, to use to make fertilizer) have migrated to cooler water in ecosystems with conditions conducive to their spawning and growth. In some cases, predators follow fish they prey upon that have migrated to cooler waters, but in other cases they find new prey to sustain them. In other situations, they may become prey for larger fish in an ecosystem. Fishing fleets adapt by following the fish they hunt into cooler waters where ideally they capture the hunted species in quantities allotted them by national and international fishery governing body regulations. If the quota system is followed, this will allow recovery of fish populations and sustainable harvesting.

Aquaculture operations that provide important supplies of food fish worldwide can adapt to warming waters by raising food fish or shellfish that will grow and multiply under the changed range of day/night temperature conditions if the fish they are farming cannot survive in the warmer waters. Aquaculturalists also have the option to move their facilities to cooler-temperature waters, but the economic feasibility of doing this has to be evaluated by a benefit to cost analysis. This analysis has to be for the time frame during which the cooler-ecosystem waters are estimated to remain stable within the framework of a time range against global warming/climate change. Another adaptation is that food fish currently being raised can be genetically engineered to be resistant to a warmer growth environment without changing their nutrition yield, growth rate, and ability to reproduce.

9.5   Adaptation to the Threat or Onset of Endemic, Epidemic, and Pandemic Disease

There are diseases that are global threats, others that put regions at risk, and yet others that menace smaller political divisions. Humans adapt to the threat of sickness in a population or a sickness itself in several ways. Scientists develop methods to eradicate a virus or bacterium health threat, or a chemical/radioactivity threat. Failing this, health professionals act to control a disease, to slow or minimize its transmission, and to apply approved therapies and support research to find therapies to treat an illness if one is transmitted. The following discussion draws strongly on the 2013 disease fact sheets put out by the World Health Organization.

9.5.1   Adapting to Global Health Threats

Vaccines provide a main line of defense against many diseases. Smallpox has been eradicated on earth by vaccination. Polio has all but been eradicated globally except for a few pockets of the disease in Pakistan, Afghanistan, and Nigeria where, in some cases, religious fundamentalists have beaten and killed health workers tasked with giving the vaccine to children, and in other cases where parents have been warned by the zealots against allowing their children to be vaccinated. Recently, 81 polio cases were diagnosed mainly in Somalia but also in Kenya. This is attributed to the fact that by 2013, 500,000 children in Somalia have not received the vaccine and are at risk from this highly contagious disease. It is also attributed to cross-border migration of infected persons into Kenya. Both governments are stepping up their vaccination programs. There were 59 cases of polio diagnosed in the rest of the world in 2012.

Measles is a global disease that can be prevented by a vaccine that is safe and cost-effective. Measles may soon reach the near-eradication stage. In 1980 and subsequent years, 2.6 million people, mainly children under 5 years of age, died from measles. Since 2000, 1 billion children were vaccinated, 225 million in 2011. By 2011, 84 % of the world's children received the measles vaccine, up from 72 % in 2000. From 2000 to 2011, deaths from measles dropped to 71 %, from 548,000 to 158,000. When the vaccination rate reaches 95 %, mainly in low-income countries, the world will have brought another disease close to elimination [6].

Seasonal influenza is a global viral illness that afflicts 3–5 million people. The sickness kills 250,000–500,000 people with severe symptoms annually. Transmission of the virus takes place when an infected individual coughs or sneezes without covering his/her mouth and releases droplets that can be inhaled by someone up to a meter away. Transmission can also be from hands carrying the virus. Seasonal influenza affects all age groups, but children less than 2 years old, people over 65, and those with complicating medical problems are most at risk. Influenza is a disease to be controlled. The principal control is by safe and effective vaccines that can prevent 70–90 % of influenza cases in healthy adults. Secondary controls are obvious for infected persons: cover the mouth when sneezing or coughing, and wash the hands frequently. The influenza vaccine is taken once annually. Because strains of the influenza virus change from year to year, adaptation is needed. The adaptation is via a vaccine that is prepared with 3 or 4 strains that scientists determine will be most common during a coming season [7].

Other types of influenza and respiratory illnesses have the potential to cause an epidemic or pandemic. They include avian flu and its strains and swine flu if the strains develop the ability for person-to-person transmission after infection, and SARS (severe acute respiratory syndrome) and Middle East respiratory syndrome (MERS) because there is human-to-human transmission of the sicknesses. To the present, the outbreaks of the animal influenza diseases have been contained by quarantining infected people during treatment and by culling flocks and herds, or if available, vaccination of healthy animals. The latter two respiratory illnesses are caused by the coronavirus, and infected people have been in isolation wards. For SARS, an illness that broke out in 2003 and spread to 24 countries, isolation of victims and treatment with antiviral drugs and steroids stopped the disease during 2004. MERS is a recent (2012/2013) illness that has been confined to Jordan, Saudi Arabia, Qatar, and the United Arab Emirates. The MERS virus has been found in camels. Infected persons are quarantined in hospitals, but an effective drug treatment is still being sought to complement the normal hospital care-afforded patients.

HIV/AIDS is a global epidemic that killed 25 million people in three decades since 1981. Worldwide, in 2011, there were 34 million people with HIV, mainly (33 million or 97 %) in sub-Saharan Africa and South/Southeast Asia. The illness is caused by the exchange of body fluids (semen, vaginal excretions, blood, breast milk) from an infected individual with an uninfected person. More than 50 % of the cases of HIV are from heterosexual activity. There is no vaccine against HIV/AIDS, no cure for it, but there is a cocktail of medicines (antiretroviral treatment) that control viral replication and allow an infected person's immune system to strengthen. This keeps the illness at bay and afflicted people in general good health and productive in their communities. In 2012, only 9.7 million (less than 30 %) of those with HIV in low and middle economies received the antiretroviral treatment. This is changing as more HIV carriers have access to antiretroviral therapy and there are more donations from economically advantaged countries to support HIV stabilization and reduction programs. The number of new cases of HIV is not exploding because more than 50 % of those infected are following protocols that reduce the transmission of the disease.

The prevention of transmission methods include access to male and female condoms, blood screening before transfusions, and needle and syringe exchange programs for sterile injections by drug users. HIV testing and education programs and HIV treatment help prevent transmission because individuals in continuous treatment have a very low probability of passing on the disease. Male circumcision reduces the infection in men by about 60 %. There is still much progress to be made because there were 2.5 million new cases of HIV in 2011, with 1.8 million of that total in sub-Saharan Africa. The HIV/AIDS is a global sickness that is slowly coming under control because of generous donations from governments and foundations in developed countries added to what low- and middle-income countries themselves provide to lower the prevalence and incidence of HIV in their populations [8].

Tuberculosis (TB) infected 8.7 million people globally in 2011, killing 1.4 million persons. It is a bacterial disease that spreads among people when infected individuals cough, sneeze, or spit, releasing bacteria into the air where they can be inhaled by others a meter away. Although TB occurs worldwide, developing countries carry the largest burden of cases and deaths (95 %). The bulk of new cases are regional in Asia (60 %) with sub-Saharan Africa reporting a large share as well with 2,600 new cases per million inhabitants. There is no vaccination for TB, but the disease can be treated and cured. The treatment is a half-year course of four antimicrobial drugs that must be taken without fail and thus requires continual supervision by healthcare personnel. More than 51 million people have been treated and cured of TB since 1995 and perhaps 20 million lives saved by following the WHO Stop TB Strategy protocols including securing adequate, sustained financing, ensuring early reliable detection and diagnosis, and providing approved treatment with a secure effective drug supply. The number of people infected with TB is declining, and from 1990 to 2011, the TB death rate dropped more than 40 %. The success in dealing with TB is muted somewhat because a strain of the bacterium that causes TB has evolved to be multidrug resistant (MDR-TB). In 2011, 310,000 cases of this variant were reported (of the 8.7 million cases worldwide), mainly from India, China, and the Russian Federation. These are treated with, but do not always respond to, the most effective anti-TB drugs. Research into new drugs to deal with this problem is ongoing [9]. There is the question of whether people visiting or immigrating from these countries should be screened before a host country issues them entry visas.

9.5.2 Adapting to Regional Disease Outbreaks

Regional illnesses threaten the health of 100s of millions of people mainly in tropical and subtropical areas and often affecting children. One of these, the guinea worm disease, is trending toward elimination, if not eradication. This is a parasitic disease caused when people swallow water contaminated with infected water fleas (microscopic copepods) carrying worm larva. The worms release, penetrate the intestines, and move through the body migrating under the skin until they emerge causing swelling and blistering. People infected with guinea worm disease cannot contribute to their communities for months. During the mid-1980s, there were 2.5 million cases mainly in 16 African nations. But attention to where the sources were so that they could be avoided and treated, and assistance in generating clean water, were adaptations that brought the number of cases down to less than 10,000 in 2007. The number of cases continued to decline and was reduced to 542 in 2012 in four African countries: South Sudan, Chad, Ethiopia, and Mali. There is no vaccine against guinea worm disease. Health officials adapt to counter this sickness in several ways. As noted above, access to clean drinking water is the best way to prevent infection. The prevention or transmission of the worms from infected individuals to healthy persons by proper treatment and hygiene and the use of the larvacide *temephos* to eliminate the parasite-infected water flea vector and other prevention protocols are important in the control and effort to eliminate/eradicate the disease [10]. The (Jimmy) Carter Institute, Atlanta, Georgia, USA, has been a principle force since 1986 in the fight to rid the world of guinea worm disease.

In tropical and subtropical regions, there are three mosquito-vectored diseases that put millions of people at risk: yellow fever, malaria, and dengue fever. Yellow fever is an endemic viral disease in tropical regions of Africa and Latin America with 200,000 cases reported annually that cause 30,000 deaths. There is no set treatment for afflicted people, but there is an adaptive preventive measure. A vaccine against yellow fever is available that is safe, affordable, and that gives lifelong immunity to the disease with one dose after 7–10 days for 95 % of the people vaccinated. When there is the onset of a yellow fever outbreak where the population lacks vaccination protection, mosquito control is an essential first step in adaptation to prevent or slowdown transmission of the yellow fever virus. Spraying insecticides to eliminate breeding sites and kill adult mosquitos is the control used during epidemics to make time for vaccination campaigns in a population and for immunity to take hold. There are limitations to the application of the yellow fever vaccine. First is that babies less than 9 months of age should not be vaccinated or, during an epidemic babies less than 6–9 months of age should not receive the vaccine. Second, pregnant women should not be vaccinated except when there is an outbreak of the disease. Third, people with a strong allergy to egg protein or those with a marked immunodeficiency or with a thymus problem should not receive the vaccine [11].

Malaria is a parasitic disease caused by the bite of an infected mosquito. There is no vaccine against malaria, but one is undergoing a clinical trial in seven African nations with results expected in 2014. A use or no use decision as a control method for malaria will be made in 2015. Promising results from an early-stage clinical trial of an unconventional vaccine prepared with live, weakened sporozoites of the malaria parasite were published in 2013. Plasmodium falciparum was given to healthy 18–45 year-old volunteers intravenously. The volunteers were grouped to receive 2–6 doses and subsequently exposed to bite by five mosquitoes carrying the parasite. None of the six that received five doses were infected with malaria. Three of the 12 that received four doses became infected, whereas 16 of the 17 that received lower doses became infected. Of 12 that received no vaccine, 11 became infected. Those that became infected were treated with malarial drugs and cured. Clearly, higher dosages give protection against infection by malaria [12]. More research and extensive clinical trials are necessary to determine how children respond to the vaccine with adjusted dosages and whether the results from early-stage trial are reproducible in larger volunteer populations. If the results of additional clinical trials go well, the hurdle of producing enough vaccine and adapting it to injection has to be faced.

Forty percent of the deaths from malaria are of African children in the Democratic Republic of Congo and Nigeria. In addition to sub-Saharan Africa, populations in Asia (especially India and the Greater Mekong region) and Latin America suffer from the disease. The effort to deal with the disease that is preventable and curable now centers on control and treatment to reduce the number of cases. In 2010, the WHO reported that there were 219 million cases and 660,000 deaths (with an uncertainty range of 490,000–836,000). In a 2012 report, researchers suggested that the number of deaths was understated and that their computer model gave a figure for 2010 almost double, 1,238,000 deaths (95 % uncertainty interval of 929,000–1,685,000) [13]. The WHO stood by its figure stating that much of the data in the cited study were based on verbal testimony of how people had died, not on laboratory diagnosis of samples. Either figure represents too many deaths from the disease and have to be reduced. Mosquito control is the adaptation that can reduce the transmission of the disease greatly. This includes personal protection by use of proper clothing and/or the application of mosquito repellent, the use of long-lasting insecticidal (pyrethroids treated) nets to kill mosquitos and prevent night-time bites, and indoor residual spraying (remains effective for months). Those people infected can be treated with oral *artemisinin* monotherapy followed by a second drug. Failure to complete the treatment as prescribed leaves parasites in a person's blood. No other antimalarial treatment is available so that parasite resistance could become a serious problem. For visitors to a malaria region, antimalarial drugs taken before, during, and after a trip can protect them from the disease. Many countries in tropical and subtropical areas have used the above-cited strategies and others to work toward the elimination of malaria. Malaria eradication is the goal of the WHO [14].

Dengue fever is a female mosquito-borne virus that infects people with an influenza-like disease in tropical and subtropical regions worldwide. The disease can kill if it evolves to severe dengue. It is endemic in Latin America and Asia where most cases now occur. Since the 1970s, the sickness has spread to more than

100 countries putting about 35 % of the world's 2013 population (2.5 billion people) at risk. Dengue fever is especially endemic to urban/semi-urban environments. Humans are the main carrier of the virus. After a mosquito bites an infected person, each subsequent bite by the infected mosquito creates another carrier. A mosquito can bite many people each time it feeds. In the Americas alone, there were 1.6 billion cases of dengue fever reported in 2010 with 49,000 being severe dengue. There is no vaccination against dengue fever, but research continues to develop one. The main treatment for afflicted persons is to keep them hydrated. Adaptation to deal with slowing or stopping the spread of dengue fever involves three main tracks in addition to spraying insecticide to kill mosquitos. The best control method to prevent the transmission of the virus is to deprive mosquitos of sites with shallow, standing water where they can lay eggs and multiply. Control can be improved if communities cover and clean water storage containers regularly, and use proven insecticides on them as necessary. Finally, individual protection such as the use of mosquito repellants and insecticide-impregnated bed nets can help reduce the incidence of dengue fever as it has with malaria [15]. Although controls are known, they are not always applied because of economics and other factors that prevent access to protection methods. The result is that the number of cases of dengue fever reported continues to grow globally. As populations increase in urban locations, the incidence of dengue fever can be expected to increase as well unless strict controls are enforced until a safe and cost-effective vaccine is developed. A positive aspect of the dengue fever problem is that recovery from one serotype of the virus gives immunity for life. However, there are four serotypes of the infectious virus so that recovery from one leaves a person susceptible to the others [15].

Chagas is another regional disease. It infects 7–8 million people annually, mostly in 21 Latin American countries. It is a parasitic illness that evolves after the bite of a blood-feeding *triatomine* bug, often on the face, where it defecates close by leaving parasite-bearing feces. Parasites access the body when the feces are inadvertently smeared into the bite, the eyes, the mouth, or any skin lesion. The parasites circulate in the blood expressing their presence as a purplish swelling of one eyelid or as a skin lesion. There are several other symptoms as well in this acute stage of the illness, but these may be absent or mild. If diagnosed early during this stage, Chagas disease is treatable. The parasite is killed with the medicines *benznodazole* and *nifurtimox* taken for 2 months. There are limitations as to who can take these medicines (e.g., not by pregnant women or people with kidney or liver problems). The untreated sickness can cause cardiac alterations and digestive problems that show up 20–40 years after an untreated infection. Chagas disease can be spread by blood transfusion and by organ transplant, making blood screening for the parasite essential before a procedure. It can also pass to a fetus from an infected woman. There is no vaccination against the illness so that control of the vector (*Triatomine* sp.) is necessary. The controls adapted by many municipalities include insecticide spraying inside a home, the use of treated bed nets, and hygiene practices that protect food, its preparation, and its storage before eating it [16]. The sickness may recur if control practices become lax. Chagas disease is spreading as populations emigrate from Latin America to northern countries. Blood screening of visitors or immigrants from the countries where Chagas is endemic may be necessary, and treatment followed by an infected individual before a host country issues an entrance visa. This would prevent the ingress and possible spread of Chagas.

9.5.3 Adaptation to Localized (National, City) Health Threats

Outbreaks of diseases in town and cities is most often caused by bacterium-contaminated water or food and poor sanitation. Sicknesses such as cholera, typhoid, and various other diarrhea types are examples of such diseases. They are all highly infectious if good hygiene practices are not followed. These diseases are endemic in many countries where populations do not have access to safe water and adequate sanitation. There are vaccinations for some of these sicknesses that may require more than one dose, but they may not be completely effective or long lasting and require revaccination at times specified by medical personnel (e.g., after 2–5 years). Otherwise, infected persons can be treated with medicines such as oral rehydration pills or antibiotics. Adaptation for prevention is easier called for than realistically available: washing hands with soap and clean water after visiting the toilet, and as noted above, access to safe water and good sanitation. Given the millions of people infected by these bacterial diseases and the hundreds of thousand that die from them annually, generally in economically disadvantages countries, there should be an expanding global priority to eliminate the disease-causing conditions, and preparedness to combat an outbreak when it is reported.

9.5.4 Planning Ahead to Stem Future Health Threats

There are important factors to consider when adopting plans to halt or meliorate the effects of health threats to people in the near and extended future. One is the climate change-driven spread of tropical and subtropical diseases discussed earlier to newly warmer and moister higher-latitude and higher-altitude zones. Another is the growth of populations mainly in tropical and subtropical regions in Africa, Asia, and Latin America. Together with this latter factor are the increasing populations and population densities in urban centers especially in the regions just cited. An additional factor to consider is whether there is accessibility to populations by healthcare workers or by people to healthcare clinics or hospitals, well-staffed and well-stocked with necessary pharmaceuticals.

Certainly, future planning has to include funding to support research to develop vaccines for diseases that do not have vaccination as an option against an illness (e.g., malaria, dengue fever) . In addition, improvement of vaccines that are available but that are not completely effective in terms of protection or the length of time they are effective should be a priority in pharmaceutical and biotechnology laboratories.

Scientists presented a fine review of the status of vaccine research from the design and development of vaccines to discussion of vaccines and infectious diseases (e.g., HIV, malaria, tuberculosis, pneumococcal disease, and influenza) [17]. They also discuss vaccines against enteric infections and viral diseases of livestock as well as vaccines against non-infectious diseases (e.g., cancer) and against chronic non-infectious diseases. Continued and repeated education classes on how to prevent the transmission of diseases and free supplies of materials that work to this end (e.g., insecticide-treated bed netting, male and female condoms) are essential to reducing the prevalence and incidence of diseases as are safe water and uncontaminated food. As new medicines or combinations of medicines are developed, tested, and found to be effective in controlling diseases, they become part of the protocol for either curing disease or controlling disease to reduce transmission while allowing persons to carry on with their lives. In these times of easy and rapid migration, one wonders whether screening of visitors or immigrants for diseases known to be endemic or active in the countries or regions from which they come should be required so as to prevent a carrier from infecting others and spreading a disease (e.g., Chagas disease, cholera, tuberculosis). This was done at airports during the SARS scare for people leaving or entering a country (e.g., China) and likely limited the transmission of the SARS virus and spread of the disease.

Preparedness for a disease outbreak, response to an outbreak, and management of resources during and post-outbreak are the keys to adapting to health threats that could affect future generations. This means developing the capability to extend the reach of health services to regions where climate change brings warmer, moister conditions to higher-latitude and higher-altitude ecosystems that are now reached by disease vectors that have expanded into these formerly cooler and drier environments as a result of global warming. Adapting to this reality and planning ahead makes it possible to deal with and stem an incipient outbreak of disease before it is transmitted and spread to the general population. This becomes essential when there is a future disease outbreak in large, dense populations in tropical and subtropical urban centers as well as those in regions warmed and humidified by climate change to subtropical and tropical settings. Remember that urban populations worldwide, especially in Africa, Asia, and Latin America, are where much of the global population growth will take place during the next few generations. Under these conditions, diseases can spread rapidly in many ways. These include from bites of vectors, by respired droplets after an infected person coughs or sneezes, and by touching surfaces bearing viruses, bacteria, or parasites. Diseases are also spread by ingestion of contaminated water and/or tainted food, and by other methods of infection transmission. Disease transmission can be checked by rapid response teams with appropriate and sufficient supplies to treat (and perhaps places to quarantine) those in the infected population.

Lastly, it must be noted that there are many other diseases in addition to those cited previously for which prevention, treatment, and cures are research priorities in laboratories worldwide. In addition, there are addiction diseases that can trigger health problems in important segments of society. These include smoking (e.g., emphysema, lung cancer), alcoholism (e.g., cirrhosis of the liver), drugs (e.g., various psychological and physical ills), and overeating (obesity, diabetes, high blood pressure). Adaptation to these health threats involves public education forums through various media outlets, counseling, and sponsored groups with their individual group meeting, and programs are assisting many in breaking from an addiction to the benefit of a healthier life. Adaptation to meet the health challenges in the past, and in contemporary times has been a slow, progressive adventure with many successes but with much yet to be done. This is the planned path for the future: meet the challenges of societal health threats, resolve many, and keep researching to resolve others.

9.6 Afterword

A special IPCC report in 2012 examines in a general way adaptation to a changing climate as a risk management approach [18]. It uses pre-planning to reduce exposure and vulnerability to extreme hazard events by preparing for them beforehand, responding to their impacts on people, structures, and infrastructure, and having in place recovery systems that can act when a danger condition eases. In this way, there will be an ability of populations to cope with future risks brought on by a changing force with which a hazard impacts a community, changes in the frequency of an occurrence, and extension of the spatial reach of its destructive power. Much of this has been discussed in the chapters of the book you are reading.

An understanding of what is being done now to adapt to the various problems society faces during the second decade of the twenty-first stimulates proposals of how to adapt to them as global conditions change in the future. To this end, the World Bank commissioned a study on the effects global warming as it increased from 0.8 °C that exists on our planet now to what can be expected if the warming reached 2 °C, a change that many scientists believe we can adapt to, and then reached 4 °C as warming continues [19]. The study centered on regions with high population growth and great susceptibility to be negatively impacted by climate changes: (1) sub-Saharan Africa where food production is at risk: (2) Southeast Asia where coastal zones and productivity are at risk; and (3) South Asia where there could be extremes of water scarcity and excess. The effects of higher temperatures from global warming and climate change included what has been discussed in previous chapters of this book: heat, drought, sea level rise, coastal zones, typhoons, flooding, river runoff, water availability, ecosystem shifts, crop yields, fishing, aquaculture, livestock, health and poverty, and tourism. Projections such as those published in the World Bank study give impetus to governments, international institutions, multinational companies, private foundations, and NGOs to think now, to invest now, and to research now for adaptations that can be realized in good time and that will provide global citizenry with a good quality of life where needed.

9.7 Epilogue

In this book, we have examined existing human populations and the problems they are experiencing in the second decade of the twenty-first century and have also considered growing populations globally and additional problems future generations will experience. We have discussed strategies on how to cope with many-faceted threats to citizens. These include how to nourish those who need food and water, how to shelter people safely from natural and anthropogenic hazards, how to provide them with healthcare, education, and employment, and how to prepare them for the evolving global warming and the physical and biological dangers that ensue from climate change. Given the present global conditions with about 14 % of our earth's population suffering from malnutrition and more than 21 % not having access to safe water, our capability of nourishing a billion and a half more people by 2035 is in question. Also problematical is our capability to provide for an additional billion people 15 years later, or a total of at least 10.3 billion people by the turn of the century, that is, if we reach those population figures or have population crashes such as from pandemics that can kill scores of millions if a disease is not immediately treatable, or an unlikely but possible nuclear conflagration that could do the same. Less likely yet is an explosion of a small asteroid or comet in the atmosphere such as happened in a poorly inhabited area of Siberia in 1908. Here, an exploding mass more than 60 m in size knocked down millions of trees in an area greater than 2,000 km^2 (close to 800 mi^2) with energy thought to be 1,000 times greater than the Hiroshima atomic bomb. Clearly, such an event could kill the population of a megacity if it were to occur.

Another question is whether national governments are economically strong enough and have the will to set priorities that adopt strategies to protect citizens from natural (e.g., earthquakes) and anthropogenic (e.g., pollution) hazards as well as from extreme weather conditions that are supported by global warming (pollution of the atmosphere) but are naturally occurring. Countries can also improve social and economic conditions by investing in health care and education for their citizens in order to form a sound and knowledgeable cadre that would be attractive to investors interested in locating a development project that would provide employment. Again, this is in question given limited national economic capabilities and the increasing numbers of people to be accommodated, especially in several developing and less developed countries in Africa, Asia, Latin America, and the Middle East.

At this point, we must ask, "What is the carrying capacity of the earth?" Have we reached it at 7 billion given the billions who are today under served in developing and less developed countries? Some scientists will answer yes, whereas others believe that advances in agriculture and technology can allow population expansion although to what point is not defined. Can countries that are poisoning their environments do a turn around to save their citizens from grief? Can they exert controls on operations that create unhealthy conditions that sicken people, lessen agricultural production, and otherwise disrupt local, regional, and global ecosystems at the expense of maintaining their GDP and increasing it? This is not the case for many nations today that do not want to acknowledge that changes tomorrow may be too late and that the future begins now. Without action now to activate programs to sustain and nurture our ecosystem earth, the future is bleak for many and bright for a few. With action now, we can strive toward an equalization in benefits for all citizens. The Intergovernmental Panel on Climate Change completed the Fifth Assessment Report April, 2014 after this book was written. "Climate Change 2014: Impacts, Adaptations, and Vulnerability." It is a comprehensive and important contribution to the challenges presented to society by climate change now and in the future and how to manage them for the good of society. The full text is online at www.ipcc.ch/report/ar5

References

1. The National Academy of Sciences and the Royal Society (2014) Climate change: evidence and causes (36 p). The National Academy Press, Washington, DC
2. Intergovernmental Panel on Climate Change (2007) Climate change 2007 (as four part report). Part 1. The physical science basis (February 2007); Part 2. Impacts, adaptation and vulnerability (April 2007); Part 3. Mitigation of climate change (May 2007); Part 4. Synthesis report (November 2007). World Meteorological Association and United Nations Environmental Programme http://ipcc.ch/pdf/assessmentreport/ar4/syr.ar4_str_spm.pdf
3. Marzeion B, Larosch AH, Hofer M (2012) Past and future sea level change from the surface mass balance of glaciers. Cryosphere 6(6):1295–1322. doi:10.5194/tc-6-1295-2012
4. Gardner AS, Moholdt G, Cogley JG, Wouters B, Arendt AA, Wahr J, Berthier E, Hock R, Pfeffer WT, Kaser G, Ligtenberg SRM, Bolch T, Sharp MJ, Hagen JO, van der Broeke MR, Paul F (2013) A reconciled estimate of glacier contributions to sea level rise, 2003–2009. Science 340:852–857. doi:10.1126/science1234532
5. Domingues C (2012) Ocean thermal expansion and its contribution to sea level rise. Ice Sheet Mass Balance Workshop, Portland, Oregon. Abstract and video at www.climate-cryosphere. org/events/2012/ISMASS/ThermalExpansion.html. Accessed 18 Mar 2014
6. World Health Organization (2013) Measles. Fact Sheet No. 286 3 p
7. World Health Organization (2009) Influenza (Seasonal). Fact Sheet No. 211 3 p
8. World Health Organization (2013) HIV/AIDS. Fact Sheet No. 360 5 p
9. World Health Organization (2013) Tuberculosis. Fact Sheet No. 104 5 p
10. World Health Organization (2013) Dracuncukiasis (guinea-worm disease). Fact Sheet No. 359 4 p
11. World Health Organization (2013) Yellow Fever. Fact Sheet No. 100 5 p
12. Seder RA et al the RC 312 study team (2013) Protection against malaria by intravenous immunization with a non-replicating sporozoite vaccine. Science. 341(6152):1359–1365. doi:10.1126/science.1241800
13. Murray CJL, Rosenfeld LC, Lim SS, Andrews KG, Foreman KJ, Haring D, Fuliman N, Naghavi M, Lozano R, Lopez AD (2012) Global mortality between 1980 and 2010: a systematic analysis. The Lancet 379(9814):413–431. doi:10.1016/50140-6736(12)60034-8
14. World Health Organization (2013) Malaria. Fact Sheet No. 94 7 p
15. World Health Organization (2013) Dengue and Severe Dengue. Fact Sheet No. 117 4 p.
16. World Health Organization (2013) Chagas disease (American trypanosomiasis). Fact Sheet No. 340 4 p

17. Greenwood B, Salisbury D, Hill AVS (2013) Vaccines and global health. Philos Trans R Soc
    B 366:2733–2742
18. Field CB, Barros B, Stocker TF, Qin D, Dokken DJ, Ebi KL, Mastrandrea MD, Mach KJ,
    Plattner G-K, Allen SK, Tignor M, Midgley PM (eds) (2012) IPCC special report summary for
    policy makers. Managing the risks of extreme events and disasters to advance climate change
    adaptation. Cambridge Univ. Press, UK and USA 20 p Full report 582 p
19. World Bank (2013) Turn down the heat: climate extremes regional impacts and the case for
    resilience. A report for the World Bank prepared by potsdam institute for climate impact
    research and climate analytics. World Bank, Washington, DC, 254 p

Index

© The Author(s) 2015
F.R. Siegel, *Countering 21st Century Social-Environmental Threats to Growing
Global Populations*, SpringerBriefs in Environmental Science,
DOI 10.1007/978-3-319-09686-5